SONGS, ROARS, AND RITUALS

LESLEY J. ROGERS

AND

GISELA KAPLAN

Songs, Roars, and Rituals

COMMUNICATION IN BIRDS, MAMMALS, AND OTHER ANIMALS

HARVARD UNIVERSITY PRESS · CAMBRIDGE, MASSACHUSETTS

Printed in the United States of America
Second printing, 2002

First Harvard University Press paperback edition, 2002

An earlier version of this book was published in 1998 by Allen & Unwin as
Not Only Roars and Rituals: Communication in Animals

Drawings by Tina Wilson

Library of Congress Cataloging-in-Publication Data

Rogers, Lesley J.
Songs, roars, and rituals : communication in birds, mammals, and other animals /
Lesley J. Rogers and Gisela Kaplan.
p. cm.
Rev. ed. of: Not only roars and rituals. 1998.
Includes bibliographical references (p.).
ISBN 0-674-00058-7 (cloth)
ISBN 0-674-00827-8 (pbk.)
1. Animal communication. I. Kaplan, Gisela. II. Rogers, Lesley J. Not only
roars and rituals. III. Title.
QL776 .R64 2000
591.59—dc21 00-025602

To the memory of Tipsy,

a dog special to us among all animals

CONTENTS

FIGURES

PREFACE

It is reasonable to ask at the beginning of this book why communication in animals interests us. What can knowledge of animal communication achieve, both in terms of understanding our own environment and in terms of our ethical position toward the natural world?

Researching animal behavior is a humbling experience. It shows how little we know about the hundreds of thousands of species that inhabit the globe and how little we know of the ways in which they communicate within and between species. How exciting it is when we think that perhaps we may have cracked another part of a code in this enormously large world of secret codes.

We are constantly discovering more about the complex capabilities of animals. No one can help being impressed by the wealth of social subtleties and complexities that individual species display. In the songs, roars, and rituals they perform, we begin to see meaning. Here is our personal wonder, pleasure, and excitement in studying animal communication. These are qualities that ultimately sustain the most enduring inquiries.

We also know only too well that new knowledge of animal behavior is needed urgently. Many species are tumbling into extinction because of direct human intervention and human mistreatment of the precious legacy of the natural world. In some cases, we do not even know why. In others, we know why, but have found few acceptable ways of coexisting with other species. Instead, we have deprived them of habitat and conditions they need to survive. Some of that mistreatment, exploitation, or coercive control may in part be based on ignorance. As the social philosopher Hannah Arendt once said, most people who "do evil do not intend to do evil." The rate of extinction of species shows that we are doing evil and, ironically, in so many instances we are doing this while actually proclaiming our liking for animals. We harm them even by assuming that they must like, react to, and be comfortable with the same things that satisfy us. This is often far from the case. Only in the twentieth century have we humans truly begun to understand that the existence of animals and their well-being is tied to ours. In turn, our well-being, at least partly, is dependent on their being allowed to maintain their lives.

In this introduction to animal communication, we attempt to provide a sympathetic but scientifically well-founded argument about a set of complex behaviors in animals. We have considered a broad range of communicative patterns in mammals and birds, and even in frogs and other species. One focus in this book is on learning to communicate so as to suggest to the reader, and to remind ourselves, that we still need to free our thinking from the legacy of Descartes and his view that the capacities of animals are purely mechanistic. One of the aims of this book, then, is to suggest that many animals are sensitive to what they do and to what we may do to them. Many may suffer at our hands and many are doomed to slide into extinction unless we can learn to respect animals in ways that leave them unfettered space to lead their lives. Although this is an introductory book on the broad issues of communication in animals, we hope very much that it is a book to be enjoyed as much by the general reader as by the student of ethology and by colleagues in the field, offering some special morsels and giving a portrait of animals consistent with our view that animals matter a great deal.

Many of the ideas that form the backbone of this book were refined in valuable discussions with our colleagues and friends, Professors Michael Cullen, Judith Blackshaw, Richard Andrew, Peter Slater, Jeannette Ward, Dietmar Todt, Allen and the late Beatrix Gardner, and also Drs. Christopher Evans, Patrice Adret, Michelle Hook-Costigan, and Jim Scanlan. We are also most grateful to our anonymous reader for Harvard University Press for excellent suggestions and to our editors Michael Fisher and Nancy Clemente.

SONGS, ROARS, AND RITUALS

Chapter One

WHAT IS COMMUNICATION?

A large flock of galahs, Australian cockatoos, is feeding on grain scattered on newly plowed soil. Hundreds of white crests, though flattened, are distinctly visible against the birds' bright pink breasts and the background. Each bird maintains a characteristic social distance from the others and the hundreds of bowed yet bobbing heads suggest complete attention to feeding—until, catching sight of an approaching farmer, one bird raises its crest and screeches. At this signal of alarm the flock takes to the air as if the decision to do so were instantaneous. A signal has been sent and its meaning interpreted reliably by each member of the flock. Communication has occurred.

An enormous elephant seal lumbers up the beach, head raised, snorting as he threatens a rival. The animals make aggressive lunges at each other, blood is drawn, and, with growling sounds, a truce is reached. One seal bows his head and moves away. Victor and loser have communicated on a matter of disputed territory and partner ownership has been decided.

These are grand spectacles of communication, but intimate and close-range contact has its own forms of more subtle communication. A mother orangutan cradles her baby of seven days on her chest as she hangs by all four limbs. She smiles, as a human might do, and the infant glances up at her. A bond has formed between mother and infant and is maintained by communication.

As these examples show, communication in animals can take many forms, and before we explore such, it is important to have a working definition of what we mean by communication in a more general sense. There are many different definitions of communication and they vary with the field in which the researcher is working. When referring to communication in humans, psychologists often restrict the concept to acts that we perform with the *intention* of altering the behavior of another person. Linguists, however, are prepared to use a broader definition of communication to include the gestures and facial expressions that we make quite unintentionally while speaking. Since the person receiving these signals perceives and interprets both the intentional and unintentional signals, they feel it is important to include both types of signaling in discussions of communication in humans.

There is no question that a large amount of communication among humans is intentional, but much unintentional signaling takes place as well. For example, in many cultures, someone giving a friendly greeting to another person raises his or her eyebrows for a moment. This facial gesture is called "eyebrow flashing." Unless we make a conscious effort to think about it, we are not aware of having performed an eyebrow flash. Even the receiver may not be aware of having seen the eyebrow flash, despite the fact that it is a very important aspect of the greeting and alters the receiver's interpretation of the words spoken at the time. As Irenaus Eibl-Eibesfeldt (1972) has demonstrated, greetings made without the eyebrow flash are interpreted as less friendly even when the spoken words are identical. People in some cultures do not eyebrow flash (most Japanese people, for instance do not do so and, as Eibel-Eibesfeldt found, they even think it is indecent), and this can create unintentional difficulties in intercultural communication. There are many other examples of what is called nonverbal communication in humans, most of which are both signaled and received unintentionally.

We can always find out what aspects of signaling by humans are intentional by asking senders exactly what message they meant to communicate and what aspects of the signal they are aware of performing. This is not possible when we are studying communication in animals. Even if an animal is sending a signal intentionally, it is very difficult for us to prove that this is so. The question of intentional versus unintentional signaling in animals is a hotly debated topic and one of major significance for the way in which we view and treat animals, but it is very difficult to study.

We must now decide on a definition of communication that will be useful in the study of communication in animals. Communication requires one individual to send a signal of some description and another individual to receive that signal and interpret its meaning (Figure 1.1).

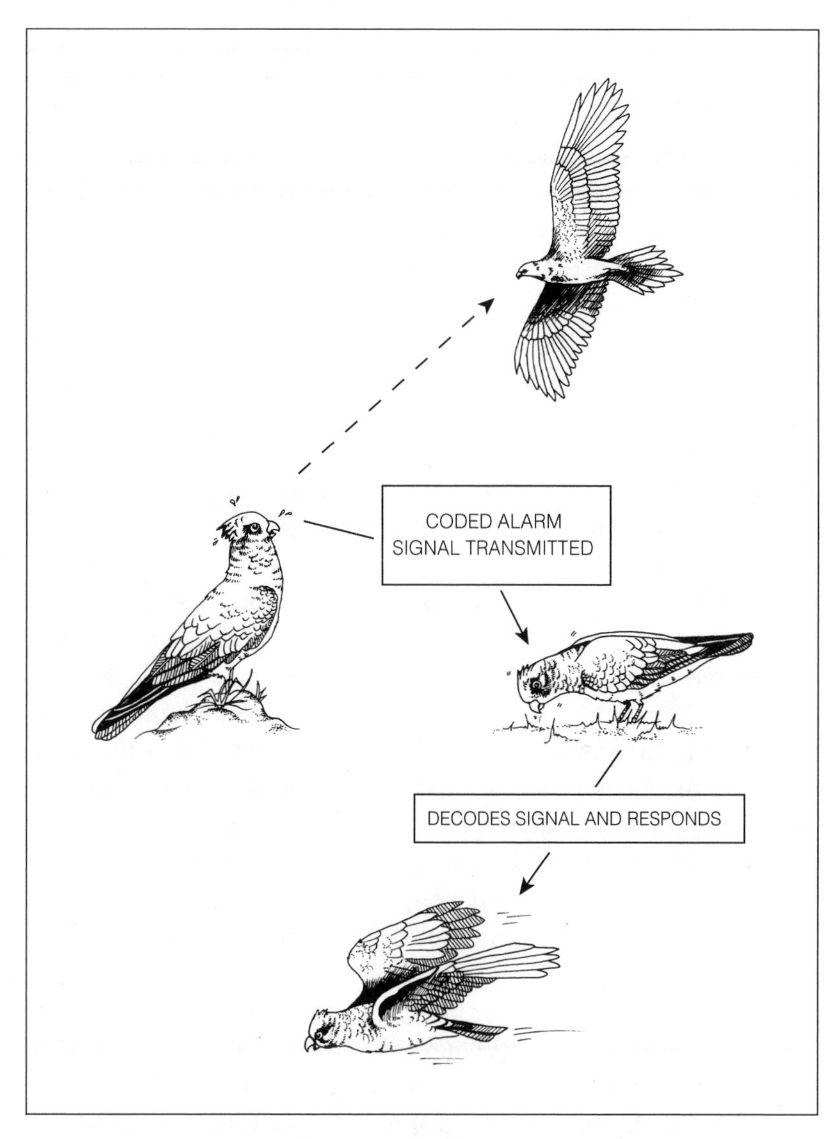

FIGURE 1.1 Sending and receiving a signal. The galah on the left has seen an eagle flying overhead and sends a coded alarm signal. This signal is transmitted through the air and detected by the galah on the right. The latter must discriminate and decode the message, which it does very rapidly, and then it responds by flying off. Note that the predator too may receive the message and respond by attacking the sender. Drawing attention to oneself is a risk of issuing a warning signal.

Biologists specify how the signal must be detected and processed by the receiver: the signal must be perceived by the receiver through one or more of the sensory systems—usually by the sense of vision, hearing, smell, or touch. In broad terms, we can say that an animal has signaled when it changes its behavior, and that communication has occurred when that signal is perceived and interpreted by at least one other animal. Of course, there are many ways in which a signal can be transmitted and many ways in which it can be received.

Communication is often seen as a way of changing another's behavior without physical force or any large expenditure of energy. In a paper published in 1972, Michael Cullen illustrated this point. He said that the command "Go jump in the lake" is a signal, whereas the push that might be delivered with it is not. The push may, in fact, convey very important information to the receiver, but the biologist views the push as physical force rather than true communication (Cullen, 1972). In this example, the verbal command is perceived by the sense of hearing, probably in conjunction with an angry facial expression perceived by the visual system; the verbal signal may change the receiver's behavior, although in this case the receiver is likely to take up a defensive posture rather than do what is commanded. The receiver might also choose to ignore the verbal signal, but could not ignore the physical force that follows the command.

The example of the alarm signal given by the galah is energy-efficient because a relatively small effort on the part of the galah that issues the alarm call leads to a large energy response: the whole flock takes flight. This signal is also time-efficient because the whole flock takes off almost instantaneously. It would be very inefficient in terms of energy and time if the signaler had to go around and physically push each member of its flock into taking off. There are, however, instances of signaling that are less efficient in energy and time. The loud roaring of red deer stags is one.

The broadest definition of a signal considers that communication occurs if any aspect of one animal's presence or behavior leads to a change in another animal. Any change in posture or other aspect of the sender is considered to be a signal. The signal, according to this definition, does not have to be specific or precisely tailored to the situation. It might merely involve a change in body posture or a slight movement of the mouth or eyes.

A somewhat narrower definition of a signal, as used by John Krebs and Nick Davies (1993), specifies that the sender (or actor) must use a specially adapted signal: a signal that has evolved to be used for communication. This is basically the same definition of signaling stated earlier by Edward O. Wilson (1975). Wilson said that communication is an action by one organism that alters the behavior pattern of another organism in a fashion that is adaptive to either one or both of the participants. The word "adaptive" is important here. Wilson said that by "adaptive" he meant that the signaling or the response, or both, have been genetically programmed by natural selection. Hence this definition confines communication to signaling and receiving that have become part of the genetic characteristics of the species. Means of communication that are learned during the individual's lifetime, and may be passed on from one generation to the next by cultural transmission, are not included in Wilson's definition of communication. Of course, genes always play some role in behavior—for example, genes determine whether we have hands or wings and that determination influences what kinds of signals we can send. But that is not what Wilson means by adaptive signaling; he means that the behavior of signaling, or the behavior of the response itself, is to a large extent controlled by genes.

STUDYING COMMUNICATION

As observers, we can tell that the signal has been perceived and interpreted only if the receiver changes its behavior in response to the signal. Therefore, to determine whether communication has taken place we may look for a change in the sender's behavior followed by a change in the receiver's behavior. Sometimes, however, the receiver of the signal may not respond. Nonresponse to a signal presents a problem to human observers of animal behavior because we have no way of knowing that communication has occurred unless the receiver responds overtly to the signal. Thus scientists who study the behavior of animals are forced to ignore signaling that the receiver ignores; they say that communication has occurred only when the behavior of the receiver is observed to change as a consequence of receiving the signal. Although we recognize that this approach means that we will overlook some of the signals that pass between animals, it is not a serious problem at present—we still have much to learn about signals between animals that do, in fact, change the behavior of the receiver. It is also likely that most signals given by animals in their natural environments cause the receiver to respond in one way or another.

Among the scientists interested in communication between animals are behavioral ecologists, who focus on special signals that have evolved to ensure the survival of the individual animal. The warning call of the galah and its ability to trigger flight is an example of this kind of communication. So too are courtship rituals that have evolved to form and maintain bonds between individuals and to ensure that mating behavior is confined to members of the same species. Many of these signals involve elaborate choreography performed in a highly stylized, stereotyped, or ritualized manner. These signals are performed with little variation between individuals and are patterns of behavior that are quite distinctive to the species. They are called displays. Displays are the same as signals, at least the most obvious ones, although usually we use the term "displays" to refer to visual signals only, not to vocal signals or signals conveyed by any of the other senses.

In 1914 Julian Huxley described the extraordinary and complex mating display of the great crested grebe *(Podiceps cristatus)*. The courting pair perform a complex ritual of precision swimming, beginning with synchronized skimming across the surface of a lake, then diving at the same time and rising together with weeds in their beaks and assuming an upright posture by treading water while they face each other. Another courting display is the male riflebird's (*Ptiloris* spp.) rhythmic opening and closing of one wing after the other; he stretches each wing over his head and bobs his head as he does so. This visual display is accompanied by sharp, explosive sounds, produced each time a wing is opened. The combined auditory and visual performance attracts the female and she responds by becoming sexually receptive. Elaborate courtship displays are performed by many species, from insects to humans.

Ethologists, who study the behavior of animals, are interested in courtship and other displays such as these but also in somewhat less ritualized signaling that could be part of any aspect of social behavior, unrelated to survival in any obvious way.

To study communication in animals we watch for changes in behavior, first by the sender (or actor) of the message and then by the receiver, sometimes referred to as the "perceiver" or the "reactor." Without the signal, there would be no change in the behavior of the receiver. But observing communication is a little more complicated than simply looking for a change in the behavior of the sender followed by a change in the behavior of the receiver because sometimes signals can actually *prevent* a change in the receiver's behavior. The receiver may, for instance, go on performing the same behavior as long as it is receiving a signal but switch to another behavior only when the signal is no longer given. The female riflebird may continue to show sexual responses as long as the male continues his courtship display but cease to do so if he stops displaying. In this case, the continued presence of the signal maintains the receiver's state of readiness.

Hence the receiver's behavior can change during the time when the sender is signaling of after the signaling ceases. We can put these two types of response by the receiver together and simply say that communication has occurred when the behavior of the receiver changes either after the signaling begins or after it stops.

Some signals are sent and received very rapidly and they cause immediate responses—the warning call is a dramatic example. Other signals act more slowly because they are not immediately detected by the receiver, especially signals that use odors. For example, marmosets, which are small monkeys of the South American rainforest, deposit scented secretions on branches; these signals are detected by other marmosets when they contact the same branches, even some time after the marmoset that deposited the message has moved away. Both short-delay and longer-delay signaling represent communication.

There is another form of longer-delay signaling in which the signal is received and processed but the receiver's behavior does not change until after a long period of delay. For example, a female wild dog, or a wolf, may signal that she is coming into estrus both by her behavior (she mounts other dogs more often) and by her odor (of secretions from the vagina and in her urine). Although these signals are received by the alpha male in the pack, he may not respond by mating until she has reached the peak of her estrus and is most likely to conceive. Delayed forms of responding to signals, such as this one, are difficult to study because it is not easy to link the sending of the signal to the receiver's change in behavior. But such signal-response delays are common among animals that attend to odors and are important aspects of communication that we humans tend to forget—compared to many species of animals, we are less aware of odors, even though they do influence our behavior.

The variations in delay time from sending the signal to changing the behavior of the receiver and in the intensity of the signal (from subtle to very obvious) mean that any definition of communication has to be quite broad. In addition, we prefer a broader definition of communication to include signaling and responding that is largely learned, and we do so for two reasons. First, learned communication can be as an important as adaptive (genetically programmed) communication. Second, in most cases of signaling in vertebrates, there is no empirical evidence to say whether or not a form of communication is largely programmed in the genes or largely learned.

CONSPICUOUS AND SUBTLE SIGNALS

Some forms of roaring or bellowing by animals require considerable energy expenditure and rather large amounts of time to signal information. The same is true of some elaborate visual displays, often used in courtship. Signaling that requires such large amounts of effort is said to be "honest signaling" because it lets the receiver know something important about the signaler, his size or physical health and strength, for example. Amotz Zahavi (1975) argues that receivers should not respond to signals unless they are honest. This would mean that honest signaling would be selected by the receivers, and so the receivers would be in control of the evolutionary process. The end result of this process of selection would be the evolution of signals that are very costly to produce; Zahavi called these "handicaps" (Zahavi and Zahavi, 1997).

The peacock's tail, used in sexual signaling, is the prime example of a handicap. It is a handicap in everyday activities, but, Zahavi says, it is precisely because the male's tail is a handicap that females prefer it to be as long and cumbersome as possible. The tail demonstrates a male's ability to survive despite the handicap, which means that he must be healthy and have other qualities that are essential for day-to-day survival. Males with longer, and more colorful, tails are likely to have "good" genes, and females will choose to mate with them. Marion Petrie, Tim Halliday, and

Carolyn Sanders (1991) have shown that peahens prefer to mate with peacocks with the largest number of eyespots on their trains. This result may explain why the apparently oversized, ornate train is likely to have evolved despite the handicap it causes the male in moving around and fleeing predators. Females prefer to mate with males that signal honestly.

The deep croaks of many species of toads are another example of honest signaling because the larger the toad the deeper the croak it can produce. The frequency of the fundamental (lowest) tone of the toad's call, therefore, signals honestly about his size and thus his potential to win in a contest (Davies and Halliday, 1978). The same can be said of the roaring of red deer stags; this vocalization requires great muscular effort and is produced most effectively by males in a good condition to fight (Clutton-Brock and Albon, 1979; also described in Bradbury and Vehrenkamp, 1998).

Some signals, however, are "dishonest," meaning that they lie about the sender's physical condition. Many signals conceal the physical state (strength or, especially, weakness) or state of health of the sender. These dishonest signals are used when the sender wants to withhold information in order to bluff another or to deceive another (this is summarized in Bradbury and Vehrencamp, 1998). The threat display of the mantis shrimp *(Gonodactylus bredini)* is a classic example of bluffing. These shrimps live in solitary burrows in coral reefs. They make use of holes in the coral, but the holes are in short supply and the shrimps compete for them. A shrimp that possesses a burrow must defend it vigorously from would-be occupiers. The resident will attack intruders that are not too much larger than itself but will flee from intruders that are much larger. There is one stage of development when the resident is very vulnerable and would be unable to defend itself should a fight ensue, and that is for the first three days after it has molted. Molting involves shedding the shell (exoskeleton) to expose a new shell underneath. This new exoskeleton is very soft and would provide no protection in a fight. The shrimp, therefore, is not in a position to attack while its new shell is soft, but it still performs the threat display to intruders as if it were able to attack. The signal is dishonest and may bluff the intruder (Adams and Caldwell, 1990).

We have discussed a number of signals that require much effort to produce. Other signals require very little effort and so cost little to produce.

Among animals we can find various degrees of economy of effort, as well as a range of signals from the very obvious to the very subtle. Conspicuous signals are more costly than less conspicuous ones. It has been suggested that the amount of conflict between the signaler and the receiver determines how conspicuous a signal will be (this is summarized in Dawkins, 1993). When both the sender and the receiver benefit from the communication taking place, inconspicuous signals should evolve. This would lead to "conspiratorial whispers" and, for example, subtle signals by which one member of a group warns the others that a predator is nearby. When there is conflict between the sender and the receiver, large and loud signals will evolve, as is the case in disputes over territory or sexual partners. In other words, a kind of coevolutionary arms race takes place; such signals become louder and louder or more and more conspicuous. The honesty of these signals is said to be ensured by their cost to the sender. There is, however, little evidence to support this idea as a general principle.

In fact, the size, strength, and duration of a signal may have little to do with sender-receiver costs and benefits and may instead be determined by the type of environment in which the signal must be sent. As we discuss further in Chapter 2, some environments cause the signal to attenuate rapidly and so demand the expenditure of large amounts of energy to send the signal in such a way that it can be detected by the receiver. The need to adapt signals to the physical environment is a most important factor in their evolution, but social factors also have an influence. The sensory systems used by the receiver to detect the signal must also evolve according to environmental requirements. They need to be attuned to detecting signals in particular environments. To overstate the case, it is no use specializing in the ability to detect high-frequency sounds in an environment in which such frequencies are not transmitted effectively.

The animal's ability to process the information that it receives and to remember the signals may also influence communication. Some signals may be designed to ensure that they are remembered. There may even be some simple formulas that assist memory and apply to a wide range of species. This might explain why a surprising number of poisonous species, from insects to toads and snakes, are colored black and yellow or black and red. Perhaps these color combinations are remembered easily by their predators. The "aim" of a species that is poisonous to eat is to signal this fact to any species that might possibly consider preying on it and to ensure that predators remember and stay away.

COMMUNICATION BETWEEN SPECIES

Most communication occurs between members of the same species (intraspecies communication), but there are occasions, as we have already seen, when one species signals to another or when one species responds to the signals of another species (interspecies communication). Communication about being poisonous to a potential predator is an example of communication between different species. Indeed, most known examples of interspecies communication involve predator-prey relationships. The potential prey signals to the predator in an attempt to deflect its attack. When cornered by a predator, the last resort of the potential prey is to try to scare off the predator by looking as big as possible, showing bright colors, or making a terrifying sound. Toads adopt a threat posture in which they puff up with air and stand high on their limbs, thereby making themselves look as large as they can.

Other strategies can be used by animals that are cornered. The sudden flash of a brightly colored signal may confuse the predator just long enough for the potential prey to get away. This tactic is used by some lizards, such as *Anolis* species, which perform push-ups and flash the dewlap when they encounter a predator such as a snake (Leal and Rodriguez, 1997). The Australian frill-necked lizard has an even more impressive display to communicate with a predator: it raises the large ruff around its neck so that, from the front, it looks many times larger; it opens its mouth to reveal the brightly colored tongue and lining of the mouth and then hisses. This striking display is usually followed by a rapid retreat, the lizard running on its hind limbs, ruff still raised. If the predator has been thrown momentarily off guard by the display, the lizard may have a slight advantage as it beats its retreat.

Another form of interspecies signaling by prey to predator is aimed at deflecting the attack to a less vulnerable part of the prey's body. The eyespots (ocelli) on the wings of some moths and butterflies may be used in this way, as was first demonstrated by David Blest (1957). When a bird is poised to attack, the moth or butterfly opens its wings to reveal the ocelli. Since birds are very interested in eyes, they tend to peck at the ocelli instead of the body of the moth or butterfly. Alternatively, revealing the ocelli may startle the bird and give the potential prey time to escape. This form of interspecies communication is so important for survival that some species have stylized the display by employing different ways of flashing the ocelli, either rhythmically or in a more static fashion.

Plovers feign injury to deflect the attention of the predator away from their offspring in the nest on the ground. As the predator approaches, the mother plover moves away from the nest in a manner that would signal she has a broken wing. This is a dramatic form of interspecies signaling. Another form of prey-to-predator signaling has been called pursuit-deterrent signaling. Tim Caro has used the term "pursuit-deterrent signalling" for communication in which the potential prey signals that it has seen the predator or that it is able to escape (Caro, 1995). The effect of the signal appears to be to stop the predator from attacking. For example, on seeing a predator, the Thomson's gazelle performs stotting (high leaps from all fours with the tail up displaying the white rump), bannertail kangaroo rats drum their feet, and swamp wrens flick their tails.

The stotting of Thomson's gazelles is an example of honest signaling, using high-energy expenditure to advertise the sender's physical prowess to the predator. It is costly both in terms of energy and in terms of the loss of valuable time for escape. A number of researchers have puzzled over this seeming contradiction. It was first thought that this was a visual signal to warn other gazelles in the herd to flee. But Zahavi (1979) suggested that stotting is, instead, directed at the predator, signaling the gazelle's physical fitness and therefore its ability to escape. C. D. Fitzgibbon and J. H. Fanshaw (1988) have provided some evidence in support of this idea. They found that a predator is more likely to attack a gazelle that stots at a low rate than one that is in better physical condition and can stot at a high rate. Stotting in the presence of a predator may thus save the life of an individual gazelle. It is therefore an "honest" signal, showing the predator what the gazelle can actually do, rather than being a form of manipulation.

The rich variety of signals between prey and predator is a manifestation of the importance of communication in survival. Communication between species may be of mutual benefit or may be aimed at enhancing the survival of the signaler over that of the receiver. Usually the outcome of signaling benefits the sender exclusively, not the receiver, but sometimes the receiver may benefit in ways that are not immediately obvious. Even though prey–predator signaling may lead the predator to abandon its pursuit, abandonment may be in the predator's interest because success is more likely to be achieved by stalking a prey that has not seen the predator, or one that is weaker and can be caught more easily.

Other forms of interspecies communication involve detection of predators but not direct signaling to the predator itself. The best example of this form of signaling is the response of vervet monkeys to the eagle alarm call of starlings living in the same area. The monkeys heed the starling's alarm signal and take cover. It is most unlikely that the starling is directing its signal to the vervet monkeys; it is trying to warn other members of its own species. The monkeys are simply able to "tune in" to the starlings' signals and exploit them.

HOW COMMUNICATION PATTERNS COME ABOUT

Many of the ritualized displays performed by animals look so bizarre to us that we wonder how they came about—and no doubt many human displays look equally bizarre to animals. Most of the various forms of signaling that are used by different species of animals have not arisen afresh in each separate species. As one species evolves into another, particular forms of signaling may be passed on, owing to the effects of both genes and learning or experience. Some signals have significance across many species, and so remain much the same over generations and in a number of species. But many signals, as they are passed from generation to generation by whatever means, go through changes that make them either more elaborate or simply different. If we examine closely related species, we can often see slight variations in a particular display and we can piece together an explanation for the spread of the display across species. Some very elaborate displays may have begun as simpler versions of the same behavioral pattern that became more elaborate as they developed and were passed on from generation to generation.

But how might signals or displays have come about in the first place? Some displays appear to have developed from movements made when the animal is getting ready to perform a particular behavior. These are known as "intention movements." Other signals may have come about when particular parts of a behavior pattern are elaborated on. Sometimes the part of the behavior pattern that is elaborated on appears to be irrelevant to the situation in which it occurs. For this reason, it has been called a "displacement" activity, although it is now debated whether the activity is really displaced or outside of context, and that is why we will use the term "displacement" within quotation marks. For example, two cocks threatening to fight may sometimes break off their aggressive displays, directed at each other, and peck at the ground with the beak closed. This "titbitting" behavior has been considered to be a "displacement" activity because it is not obviously relevant to the display of aggression. But those who came to this conclusion might merely have been unable to interpret the animals' behavior accurately enough to know what it really means. It could be relevant and observers simply cannot see that this is so, as Marian Dawkins said in her book *Unravelling Animal Behaviour*, published in 1986.

Another kind of behavior that has been referred to as "displacement" behavior in many contexts is grooming or preening. A cat that is eager to be fed but cannot persuade its owner to open the refrigerator may suddenly stop meowing and rubbing its owner's legs and switch to licking itself; or a bird that is unsure whether to eat a prey animal that it has not seen before may switch its attention away from the prey and preen itself briefly. In both these examples, the act of grooming or preening appears to be irrelevant to the main theme of the behavior pattern in which it occurs, but we may see it as irrelevant behavior only because we are ignorant of its function or purpose. It may therefore be better to refer to such examples as "redirected" behavior rather than "displacement" behavior. The performance of such redirected behavior may be observed and interpreted by another animal, in which case it serves as a signal.

Both intention movements and redirected behavior may be modified to signal to other animals. In addition, the physical and behavioral adjustments that animals make to regulate their physiological functions— to maintain body temperature within the correct range, for example— may be used to signal. We will discuss each of these in more detail later on. The point to stress here is that many elaborate displays appear to have evolved from simple behaviors that animals perform in their everyday life. These simple behaviors may also signal in subtle ways, but they become signals that are more obvious to the human observer when they have been exaggerated and so have become ritualized. Although the examples we consider are mostly the more exaggerated signals, we recognize that the less obvious signals may be just as important.

INTENTION MOVEMENTS

First we must distinguish between intention movements, which we are discussing in this chapter, and intentional signaling. In Chapter 3 we will discuss whether animals merely emit information that signals their emotional state or deliver planned communication in which signals are sent after the animal has made a decision about the context and other factors important at the time. In such cases, we use the term "intentional" to refer to the state of mind of the signaler, or at least to the cognitive processes involved in signaling. In this chapter, we use the term "intention movements" as ethologists do, to refer to those behaviors performed in preparation for an activity. Such behaviors signal what the animal is about to do next but they do not, in themselves, tell us anything about whether the animal thinks about performing them as opposed to performing them uncontrollably and without any form of thought. It is possible that the more ritualized these behaviors become—and so the more obvious they are as signals—the more likely it is that they will be performed with at least some of the intentionality we refer to in Chapter 3, but there has been virtually no research on this topic.

The first example of an intention movement that we will consider is the preparation for flight in birds. Before they take off into flight, many birds crouch, raise the tail, and withdraw the head (Figure 1.2A) and then stretch the body in the direction of the intended flight (Figure 1.2B). A bird may adopt this posture several times before it takes off, and thereby it signals to other members of the flock that it is about to fly. It has been observed that a pigeon does not usually disturb the other members of its flock if it performs flight-intention movements before taking off but that, if it flies off suddenly without these intention movements, the whole of the flock is likely to take to the air. In short, flight-intention movements signal that the individual is about to fly but should not be followed by the rest of the flock, whereas taking flight without prior flight-intention movements signals alarm and the whole flock takes off.

Richard Andrew (1956) studied in detail the flight-intention move-

FIGURE 1.2 Flight-intention postures and displays. A and B: Flight-intention movements. Crouching and raising the tail (A) is followed by lowering the tail and stretching the head and neck in the intended direction of flight (B), and these movements may be repeated several times before the bird takes off. C and D: Displays of the green heron *(Butorides virescens)* that incorporate aspects of flight-intention movements. C shows the forward threat display adopted in territorial defense, and D shows the stretch display performed as a prelude to mating. E: The head and tail-up posture of the kookaburra *(Dacelo gigas)* adopted when the bird is making its territorial laughing call. (After McFarland, 1985, and Smith, 1977.)

ments of certain species of birds and the involvement of head bobbing and tail flicking in many social displays. These aspects of the flight-intention movements have been incorporated into signaling patterns (displays). In some cases the meaning of the signal appears to be quite removed from the original intention movement. For example, the American green heron signals pair formation and courtship by adopting a posture in which the head is withdrawn, the beak held in the air, and the feathers on the head sleeked down (Figure 1.2D). This appears to be a modified flight-intention movement, and it contrasts with the species' aggressive display in which the beak is pointed forward, the feathers are ruffled, and the tail is vibrated (Figure 1.2C). The aggressive display also has elements of flight-intention movements but it includes aiming of the bird's weapon (the beak) at its opponent. There are even elements of modified flight-intention movement in the posture of the kookaburra *(Dacelo gigas)* when it makes its territorial call, which sounds like laughing.

Similarly, a gull about to attack stretches its neck out horizontally and directs its beak at its opponent, as Niko Tinbergen (1960, 1965) so clearly described for herring gulls. Tinbergen also described the upright threat posture of the herring gull, in which the neck is stretched upward and the head pointed downward (Figure 1.3A). Having adopted this posture, the gull struts toward its opponent. The positioning of the head and neck is exactly the posture that a gull adopts in circumstances in which it actually pecks its opponent. When the bird uses this posture as a threat display, but without actually pecking, it is a strong signal that the gull is about to attack. In this case, the intention movement of pecking has been used as a signal to display aggression.

Displaying of weapons is also characteristic of aggressive or threat displays in mammals. When a dog is about to attack, for example, it bares its teeth in preparation for biting. Bared teeth have become a display that signals aggression in many mammals. But we must add that the bared-teeth display is accompanied by changes in the eyes, ears, and body posture of the dog. Only by taking all these features into account can we accurately interpret the meaning of the bared-teeth display (see Andrew, 1965). A dog with its teeth bared, eyes open wide, ears erect, and tail up in a confident posture is threatening (Figure 1.3B), but one with teeth bared, eyes almost closed, ears flattened, and tail down between its legs is

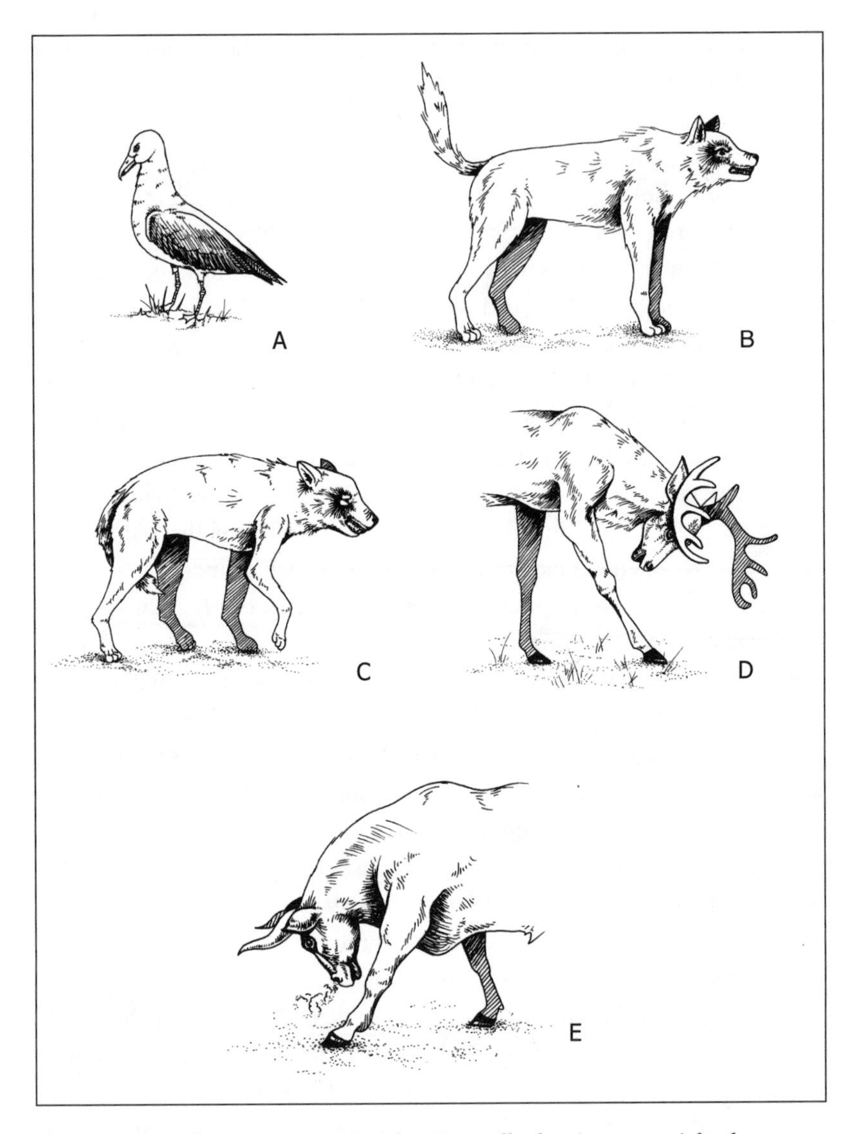

FIGURE 1.3 Threat postures. A: A herring gull adopting an upright threat posture (drawn from a photograph by Tinbergen, 1960). B: A dog with teeth bared, eyes open wide, ears forward, and tail up in a confident threat posture. C: A dog with teeth bared, eyes almost closed, ears flattened, and tail lowered in a defensive threat posture, indicating a high level of fear. D: A deer threatening to attack by displaying its weapons. E: A bull threatening in the same way.

afraid (Figure 1.3C), and will flee unless it is cornered, when it will attack. The latter display indicates that the dog is feeling a mixture of aggression and fear.

There are many other examples in which the showing of weapons signals that the animal is about to attack. A deer (Figure 1.3D) or a bull (Figure 1.3E) about to attack lowers its head and orients its horns at the object of aggression. These are all postures adopted in preparation for an actual attack, but in the display itself they do not actually lead to an attack on the opponent. Instead they are examples of intention movements being used to signal the possibility of aggression. Most such displays of aggression end with one animal backing off, thereby avoiding serious injury.

"DISPLACEMENT," OR REDIRECTED BEHAVIOR

We gave the example of titbitting in cocks as "displacement" behavior, on the grounds that when it interrupts aggressive displaying, it appears to be unrelated to the main purpose of the display. The cock pecks at small objects on the ground without picking them up, and thus the behavior does not appear to be related to the goal of feeding either. It may, however, have a function in relieving tension briefly and thus reducing the chance that an actual attack will occur. Cocks often threaten each other at the borders of their territories, and in such cases each animal experiences a conflict between approaching (and being attacked) and fleeing. "Displacement," or redirected, activities are often engaged in when an animal experiences conflict. Titbitting is one example, and by observing this behavior, another animal might recognize the position of the border of the territory. In other words, the "displacement" behavior might become a signal. Of course, once it has become a signal it can no longer be called a "displacement" activity because it has a genuine function related to the context in which it occurs. Taken together with our earlier point about the so-called irrelevance of "displacement" behaviors being merely a matter of our inability to understand them, this consideration makes the term "displacement" behavior very problematic indeed. Nevertheless, despite objections to the term, these *apparently* irrelevant behaviors do exist and elaboration upon them does appear to explain some forms of signaling.

"Displacement" preening is one of the most often cited cases. When such preening or grooming occurs, it may signal that the animal is in a state of conflict. We mentioned the cat that grooms itself when prevented from obtaining food. This is a form of approach-withdrawal conflict. The cat decides to lick itself instead of either approaching or withdrawing. Courtship behavior often involves conflict about whether to approach or withdraw, and this might explain why preening or grooming often occurs in courtship displays. In 1941 the Austrian ethologist Konrad Lorenz described preening as a feature of courtship displays in ducks (see also Lorenz, 1965). Shelducks *(Tadorna tadorna)* turn their heads to preen feathers on the back or wings during courtship. This is considered to be a displacement activity. The mallard duck *(Anas platyrhynchos)* does likewise but restricts its preening to brightly colored feathers on the wing. Preening reveals the feathers that have been specialized as part of the display. In the mallard the preening has become stylized, or ritualized. In the garganey duck *(Anas querquedula)* the ritualization is even greater; this species rubs its beak on specialized blue feathers on the outside of the wings in a way that mimics preening, although preening does not actually take place. The courtship display of the mandarin duck *(Aix galericulata)* has the greatest degree of ritualization of all these species of ducks. The mandarin duck does not actually preen but mimics the preening of two specialized secondary feathers that project up from its wings by touching them once only with the beak. The ritualization of the preening motion, and also the specialized feathers themselves, are an elaboration that appears to have evolved from the simple act of preening. The display is enhanced by the enlargement of the feathers that the duck touches with its beak and their rust-red color, which makes them stand out from the other feathers. The duck also raises a crest on the back of the head, increasing the ritualization of the behavior. In other words, the mandarin duck performs an exaggerated and highly specific display.

Differences between species in courtship displays may help to isolate closely related species from one another in situations where they live in overlapping territories. Through the use of different courtship signals, cross-breeding may be prevented. No confusion between species occurs as long as the signals indicating a willingness to mate are different for each species. In cases where species-specific mating signals occur, there is an intimate relationship between the evolution of a species and its communication signals.

In time, what might have begun as "displacement" (or redirected) preening has become a more specific signal employed in courtship displays. The example of courtship preening in ducks shows how signals might evolve, although it remains possible that learning has an essential role in establishing the final pattern of the display. Many other examples of ritualization of displacement behaviors have been described. Feeding of the female by the male is an aspect of courtship in many species, from budgerigars to gulls. The female begs for food much as the young do. When the male feeds the female during courtship, his provision of food is considered to be "displacement" behavior because it is not his goal at the time—his goal is to mate with her. In some species of birds, the behavior is ritualized by the giving of gifts in courtship. Like preening during courtship, feeding is part of the signaling process, indicating the strength of bonding and that the male will provide food for the female and offspring. Gift giving and feeding are of course also rituals in the courtship behavior of humans.

AUTONOMIC RESPONSES USED AS SIGNALS

Some of the behavioral and physical adjustments that animals must make to maintain their physiological state are also used to signal. These are known as autonomic responses because they are controlled by the autonomic nervous system. In humans, we know that these responses occur automatically, without conscious control. For example, in cases of extreme fear, the hair on our bodies is raised in preparation for cooling, necessary if we need to flee. Other autonomic responses occur also, but here we are interested in this particular one because raising of the hair is also common in other mammals. This fluffing of the hair, called piloerection, functions as a signal of fear in some species. In marmosets (*Callithrix* sp.) and tamarins (*Saguinus* sp.) piloerection in the form of fluffing of the tail signals fear and indicates that the monkey is more likely to flee than approach.

Richard Andrew (1972) was the first to point out the importance of autonomic responses in displays. He also noted that the raising and lowering of the feathers for autonomic control of body temperature has become a feature of many avian displays. Feather raising is quite difficult to interpret because slight raising (fluffing) of the feathers encloses air around the body and provides insulation for heat loss, whereas further raising causes ruffling of the feathers so that their tips no longer touch each other and consequently heat is lost because air is no longer trapped around the body. Ruffling often occurs in aggressive displays, when cooling might be needed, and fluffing often occurs when a bird is quiet and submissive. Laughing gulls, for example, perform an aggressive display in which they lower the head and jerk it rhythmically while making a deep call and ruffling the feathers. Galahs also indicate threat or aggression by ruffling the feathers (Figure 1.4).

Sleeking of the feathers is another way by which birds increase heat loss because it also reduces the amount of air trapped around the body. Feather sleeking occurs in the aggressive displays of some species. It appears commonly in states of high arousal, and it has also become incorporated into the camouflage posture of tawny frogmouths when a predator is near (see Figure 4.1). Whether feather sleeking is used as a signal between tawny frogmouths is unknown.

Urination and defecation are other autonomic responses that occur in a state of high arousal, evoked by very frightening stimuli. Not surprisingly, therefore, urination and defecation are used by some animals as part of fear or threat displays. Tawny frogmouths often turn and spray their extremely pungent feces at a predator approaching from below. When bushbabies (galagos, lower primates) mob a predator they frequently urinate on their hands and rub the urine on their bodies while emitting warning vocalizations and adopting threatening postures. The autonomic responses have become incorporated into the threat display.

The autonomic nervous system also controls the constriction and dilation of the pupils in the eyes, as we will discuss in Chapter 3. Pupil size may change according to the emotional state of the individual. The individual is not conscious of the change of pupil size but the observers, although not usually doing so consciously, assess and use this information. This is an aspect of autonomic function that is used involuntarily in communication in humans as well as in other species. In humans, dilated pupils give the face a more seductive appearance, and women used to put drops of the antimuscarinic drug belladonna in their eyes to dilate the

FIGURE 1.4 Feather ruffling in a galah *(Cacatua roseicapilla)*. A: Sleek posture. B: Feathers ruffled in an aggressive-threat posture. Both photographs are of the same individual. Note that in B the body feathers are raised and the feathers on the cheeks are elevated to an almost horizontal position, making the bird appear much larger than it is. (Photographs by G. Kaplan.)

pupils. They did this at the expense of being able to see clearly—the drug also paralyzes the muscle that controls the focus of the lens in the eye. In this case, signaling must have been seen as more important than receiving signals.

WHY SIGNALS BECOME RITUALIZED

A ritualized signal is one that is exaggerated, stereotyped, and usually repeated over and over. Quite obviously, a stereotyped signal states its point clearly and ritualization ensures that the signal is not easily confused with any other signal. This may be advantageous in itself, but there may be another reason why signals become stereotyped. As Desmond Morris suggested in 1957, ritualized signals are so stylized that they give away less information about the internal state of the sender than signals that are simpler intention movements, "displacement" behaviors, or autonomic responses. The last three types of signal convey information about the emotional state of the sender or indicate whether the sender is uncertain whether to attack or flee. Ritualized signals tend to conceal information about the sender's emotional state. In a sense, ritualization involves a loss of detailed information about the sender. As Morris suggests, ritualization may have come about precisely because it conceals this kind of information. Ritualization may be the signal-senders' way of manipulating the receiver without giving away too much information about themselves.

If it is the case that the sender is attempting to manipulate the receiver, the receiver might attempt to ignore the sender. The result might be increased ritualization by the sender, then increased ignoring by the receiver, and so on. This hypothesis is called the "arms race" explanation for ritualization. It differs from another hypothesis holding that ritualization came about to avoid signal confusion. The arms race hypothesis is, in fact, a far more beguiling view of animal communication than the one postulating avoidance of signal confusion. It seems to appeal to people accustomed to a pervasive advertising culture manipulating us all. Despite this appeal, there is no proof as to which hypothesis, if either, is correct. We note that, so far, researchers have been concerned with studying the conspicuous, ritualized signals that animals send. More attention to the subtle, quieter, and less exaggerated signals may change our views on the reasons for, and the evolution of, all signals.

CONCLUSION

The terms "communication" and "signaling" are quite interchangeable. Animals indulge in a great deal of communication about a wide range of matters. Their social life depends on communication. Communication occurs in even the simplest organisms that interact with each other, even if it is only for mating. In more complex organisms, a rich variety of communication occurs, making use of all the sensory systems and ranging from simple signals to complex ritualized displays. Some of these displays may be to a large extent the result of natural selection but even these signals may involve some learning. Other signals may be acquired largely by learning and be passed on from generation to generation as a form of culture. In some social situations and in certain environments "conspiratorial whispers" are the most effective form of communication, but in other social situations and environments the most effective communication involves the expenditure of a great deal of energy. The variety of signals, as well as responses, is enormous and fascinating.

Chapter Two

SIGNALS AND SENSORY PERCEPTION

Animals have a number of different senses and they make use of them all when they communicate. As humans, we are aware of the senses of vision, hearing (audition), touch (tactile sensation), taste (gustation), and smell (olfaction). We receive signals in all these sensory modalities but we are most aware of the visual and auditory ones. Language uses sound and is processed by the auditory system, but in most circumstances it is accompanied by visual signals and sometimes tactile signals as well. We are less aware of olfactory signals, and gustatory signals rarely reach our consciousness. Other species exploit these senses to a far greater extent.

The fact that we are less conscious of communication by odors or taste than by audition and vision does not mean that this form of communication is absent in humans. In his book *The Scented Ape* Michael Stoddart suggests that humans may be specialists in odor communication and that it influences our behavior much more than we think (Stoddart, 1990). Like most other mammals, we have specialized glands for releasing scents, and although we are far less able to detect very low concentrations of odors than members of many other species, such as dogs, we are quite good at discriminating between odors. We may use this ability to communicate among ourselves, but if we do, much of that communication goes on unintentionally and without our conscious awareness.

CHEMOSIGNALS

Marmosets deposit scents on trees and do so by rubbing the branches with the secretory glands on their chests or around their genitals. These odor, or olfactory, messages are called chemosignals. Some chemosignals indicate the general whereabouts of a species even when the individual who deposited the scent mark is not in the immediate vicinity. Other scent marks can signal the identity of the species, the identity of the individual, or its sex and social position. Marmosets deposit different odors that signal each of these very important social markers. All this complex information is communicated as smells, to be detected by the olfactory system.

Lemurs *(Lemur catta)* make use of the scent glands on their wrists, which they rub on their long, striped tails for olfactory communication. Alison Jolly has described "stink fights" in which a number of animals gather together on the ground with their tails raised and "throw" odors at each other by moving around and waving their tails back and forth over their heads (Jolly, 1966).

The sense of olfaction is one of the major senses of many aquatic species. Eels are extremely sensitive to very low concentrations of chemicals in the water, and it is thought that they use their sense of smell to return to the stream of their birth from miles away at sea. They may also use this sense to communicate with each other. A number of species of fish (minnows, catfish, sucker fish, and darters, for instance) release into the water an alarm substance from specialized cells in the skin whenever they incur even minor damage. Other members of the same species detect the alarm substance using their sense of smell and respond with typical fright reactions. Schooling fish, such as minnows, aggregate and swim away from the source of the alarm substance. Other more solitary fish sink to the bottom of the water and remain motionless; yet others swim to the surface and may even jump out of the water. In these species the sense of olfaction is critical for survival and plays a greater role than it does in humans.

ELECTRICAL SIGNALS

Some species have sensory abilities that humans lack, or do not use to any known degree. The ability to sense weak electrical fields is one of these. Electric fish (Gymnotidae and Mormyridae) have organs in their tails that send out pulses of electricity (up to 300 pulses per second) and these are used for navigation and detecting prey as well as for social signaling (Bullock and Heiligenberg, 1986). In some species, male and female fish pulse at different frequencies. The frequency of electrical pulses can indicate the sex of the fish and also its dominance in the social structure. Some species can vary the pulse rate to communicate. If there is a meeting between two fish sending out electrical signals at similar frequencies, either one or both will change the frequency of pulsing to avoid jamming the other's signals. Because the fish also use the electrical pulses for navigation, there would be confusion if they could not distinguish their own pulses from those of others. Dominant fish do not shift their frequency of pulsing to avoid jamming the transmission of another fish—that is up to the subordinate ones.

The Australian platypus *(Ornithorhynchos anatinus)* can detect electrical signals and uses this ability to locate its prey under water. The sensory organs used for this are located around the tip of the bill and enable the platypus to detect the electrical waves produced by the contracting muscles of its prey. As far as we know, however, the platypus has no specialized organ by which it can produce electrical signals itself, and so it is unlikely that it uses electrical signaling as a means of communicating with other members of its own species. Nevertheless, it is possible that a platypus can detect the electrical discharges generated when another platypus contracts its muscles during movement and can exploit this as a means of communication. This has yet to be studied.

TASTE

Taste is another sense that is used for communication, particularly by animals living in water. It is also used by many species of mammals, often in conjunction with olfaction. Cats, for example, deposit urine to mark territory. Other cats will approach the deposit of urine and sniff it, or even lick it or touch it with the upper lip and then lick the urine from the lip. Once on the tongue the urine can be tasted by means of receptors on the tongue and roof of the mouth. The urine contains substances characteristic of the cat from which they came, and the decay of these substances also indicates how long ago the animal was in the area. By smelling and tasting the urine the receiver can also tell whether the cat is ready to mate. In other words, the taste and odor convey information about the hormonal condition of the depositing cat, and this is possible because modified products of estrogen and testosterone are secreted into the urine.

Sheep and goats acquire similar information about the hormonal condition of the female, but they taste and smell the urine by licking it from the anogenital region (the area around the anus and genitals).

AUDITION

Animals also communicate by using a rich variety of auditory signals. Bird vocalizations are astounding in their variety, and most are within the hearing range of humans. Likewise, most mammals produce vocalizations that humans can hear, but some species also make ultrasonic calls. Although the calls of rats and mice are partially audible to us, most of their vocalizations are too high pitched for us to hear. If you walk into a laboratory full of rats in cages, you will hear a range of squeals and snorts, but the actual sounds are far more intense than you can perceive. This cacophony of vocalizations can be made audible to the human ear by recording them on a tape recorder capable of picking up ultrasound and then playing back the tape at a much slower speed so that the sounds are pitched at lower frequencies within our range of hearing.

Most rodents use ultrasounds to communicate. Mothers recognize their pups by the ultrasound that the pups produce, and they will retrieve their pups when they make ultrasonic distress calls. Adults also communicate with each other by ultrasounds. Some primates, such as marmosets, are able to vocalize in ultrasound, and bats specialize in it.

Certain communication sounds made by animals are quite pure in tonal quality, particularly those in the songs of birds. The two most beautiful examples of the musical songs of birds are perhaps the songs of the European nightingale *(Luscinia megarhynchos)* and the Australian magpie *(Gymnorhina tibicen)*. By transcribing the sound into a visual pattern, called a sonogram or sound spectrogram, we can see what these songs look like and study them in detail (see Figure 2.1). The sound pitch, or frequency, is plotted on the vertical axis (Y axis) and time is plotted along the horizontal axis (X axis). The loudness, or intensity, of each part of the sound is indicated by the darkness of the marks plotted. A pure tone is a single frequency at any one time, although this frequency may change. Other tones have a fundamental (lowest frequency) with harmonic overtones that appear as stripes above the fundamental. The absence of overtones gives the song a characteristic pure, whistle-like quality, as illustrated by the song of the Australian pied butcherbird *(Cracticus nigrogularis)*, and harmonic overtones make the song sound rich and musical (Figure 2.1). Compare these song structures with the broad band structure of rasping calls, in which the sound energy is spread across the frequencies, and referred to as "noise" (Figure 2.2).

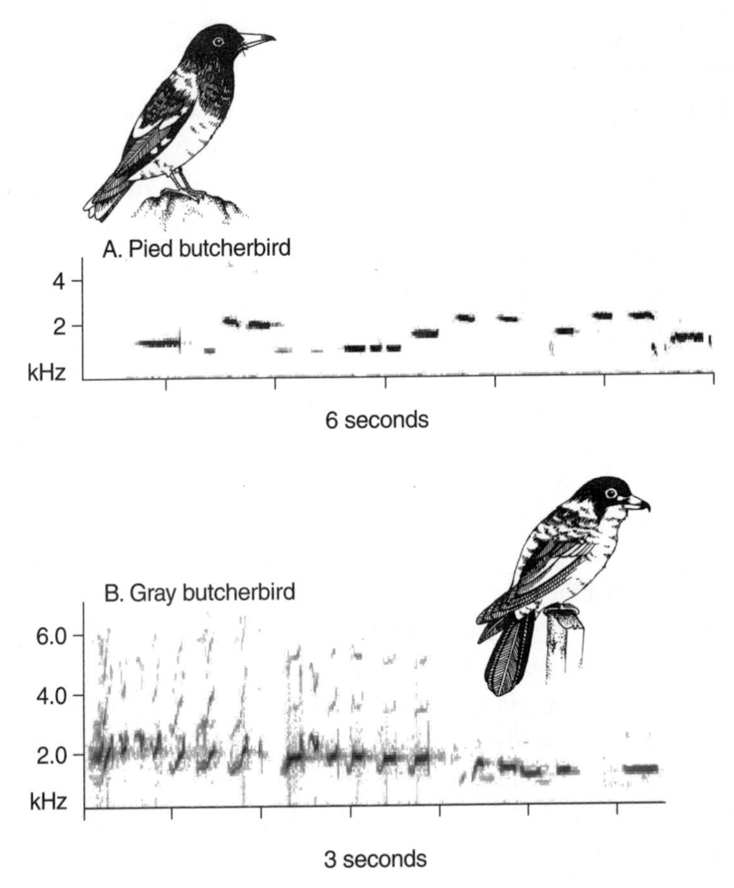

FIGURE 2.1 Sound spectrograms of butcherbird songs. A: The song of the pied butcherbird *(Cracticus nigrogularis)*. Note the pure tones. B: The song of the gray butcherbird *(Cracticus torquatus)*. Note the beginning of the song, where the syllables have loud fundamental frequencies (the dark marks at the bottom, approximately 200 kHz). Note also, above the fundamentals, the overtones (represented by marks at approximately 308 and 505 kHZ in the second section of the song). The last third of the song consists of pure tones much like those of the pied butcherbird. (Sound spectrograms produced by G. Kaplan.)

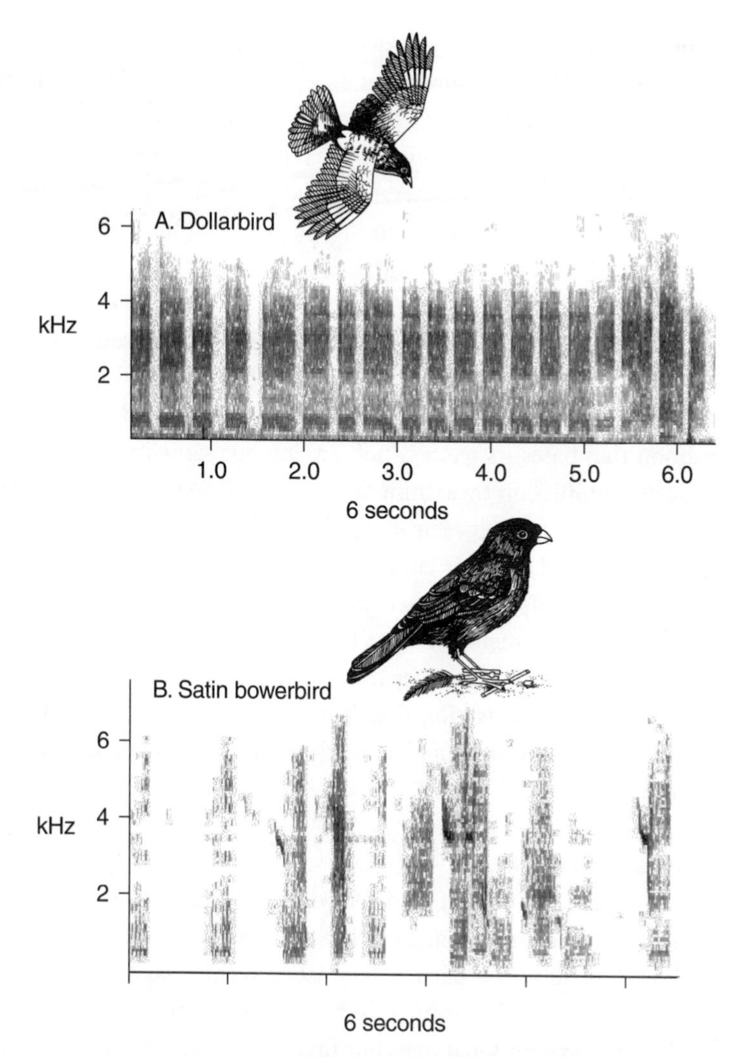

FIGURE 2.2 The sound spectrograms of noisy calls. A: The distress call of a dollarbird *(Eurystomus orientalis)*. Note that there are some overtones but that there is also a broad spread of frequencies across the range up to 5 kHz, called noise. B: Rasps made by a satin bowerbird *(Ptilonorhynchus violaceus)*. These calls have mostly a broad band of frequencies but there are some loud tones (black marks). In addition to the vocalizations presented here, the satin bowerbird produces a song that is very musical. (Sound spectrograms produced by G. Kaplan.)

Despite the fact that noisy birds do exist and that even songbirds produce some noisy calls, many birdsongs have tonality (musical sounds). By comparison, many vocalizations produced by mammals are noisy, although there are many exceptions, such as the calls of gibbons and certain calls made by squirrel monkeys, macaque monkeys, and even chimpanzees. We give further examples in the chapters that follow.

VISION

Communication using the sense of vision is widespread among animal species. These signals are made by moving limbs in certain ways (gesturing), by adopting certain body postures, or by creating facial expressions. The baboon that bares its teeth is not smiling but signaling its anger. A dog indicates submission by arching its back and lowering its tail, sometimes wrapping the tail between its hind legs (Figure 1.3). A lizard that bobs its head up and down is either signaling ownership of territory or performing a courtship ritual. There are many examples of visual signaling in animals and we mention more later on. On some occasions visual signals are used alone, but at other times they occur in conjunction with vocalizations or communication signals involving the other senses. Combined signaling with more than one sense may ensure that the receiver gets the right message.

THE BEST SIGNAL

To make sure that a signal can be detected by the receiver it is important to choose the best form of signaling for the environment. Next we will discuss how signaling in a particular sensory system has to be adapted to the habitat in which the animal lives, and we will do that by discussing each sensory system separately, but first we should say that most signals use more than one sensory system at the same time. They are thus multimodal (Partan and Marler, 1999). By stimulating more than one sensory system at once, the sender captures the attention of the receiver, and this seems to be true for humans, primates, birds, and insects as well. The begging signals of nestling reed warblers *(Acrocephalus sciraceus)*, for example, are both visual and vocal. The nestlings open their beaks wide to reveal the brighly colored skin inside and simultaneously they produce vocalizations. The parents respond to the combined effects of stimulation in these two sensory modalities by adjusting the rates at which they provide food. The two signals together provide more accurate information about the nestling's state of hunger than either of the signals does separately. In fact, the common cuckoo nestling tunes into this parent-offspring communication and exploits it to obtain food from the reed warbler parents (Kilner, Noble, and Davies, 1999).

Over a Noisy Background

When the famous musician Paganini played with an orchestra he made sure that the sounds of his violin would be heard above the orchestra by tuning it a quarter of a tone higher than the other instruments. This sharpening of pitch was not enough for his playing to sound out of tune but it did allow his notes to penetrate above the background harmony of the orchestra. This technique is exploited by some animals that need to signal over loud background noise. There is an insect in the rainforests of southeast Asia that we call "the chainsaw insect" because its penetrating whine is heard above the cacophony of other insects as a sharp, higher frequency.

Being seen as well as heard is a special problem in forests because only certain visual signals can be seen easily in the dappled and varying light of dense forest and against the background of complex patterns formed by vines, trunks, and leaves. If you wanted to send a visual signal in a "noisy" visual environment, it would be pointless to display stripes or patches of black and white. Zebras are actually camouflaged by their black and white stripes when they stand still in dappled light beside or under bushes. Black and white patterning makes effective camouflage, and the receiver would be unlikely to detect a signal that depended on the use of such patterns unless it involved movements that made the patterns more visible. It is not surprising therefore that many species of the forest are brightly colored on those parts of their bodies that are used to send visual signals.

At Dawn and Dusk

At dawn and dusk the light is purplish, and animals that forage for their food at these times are better able to avoid predators if their coloration blends in with a background illuminated by purplish light, made up of the longer (red) wavelengths together with the shorter (blue and violet) wavelengths. The middle-of-the-range wavelengths (yellow and green) are missing in purple light, and so animals with these colors would appear dark in such light. This means that the best way to signal in these conditions would be to use blue, red, or purple for brightness and yellow or green for contrast. Galahs are most active at dawn and dusk, and at these times, when they tend to feed on the ground, their grey backs camouflage them from aerial predators. Flocks of galahs are also quite difficult to see against the evening or dawn sky until the flock turns suddenly and there is a bright flash of the rose-colored breasts, a sight never forgotten. This quick turn may be some form of social signal related to group cohesion, but there is no proof that this is so.

In the Rainforest

The daytime quality of light in forests varies with the density of the vegetation, the angle of the sun, and the amount of cloud in the sky, as shown in detail by John Endler (1993). Both animals and plants have different appearances in these various lighting conditions. A color or pattern that is relatively indistinct in one kind of light may be quite conspicuous in another.

In the varied and constantly changing light environment of the forest, an animal must be able to send visual signals to members of its own species and at the same time avoid being detected by predators. An animal can hide from predators by choosing the light environment in which its pattern is least visible. This may require moving to different parts of the forest at different times of the day or under different weather conditions, or it may be achieved by changing color according to the changing light conditions. Many species of amphibians (frogs and toads) and reptiles (lizards and snakes) are able to change their color patterns to camouflage themselves. Some also signal by changing color. The chameleon lizard has the most striking ability to do this. Some chameleon species can change from a rather dull appearance to a full riot of carnival colors in seconds. By this means they signal their level of aggression or readiness to mate.

Other species take into account the changing conditions of light by performing their visual displays only when the light is favorable. A male bird of paradise may put himself in the limelight by displaying his spectacular plumage in the best stage-setting to attract a female. Certain butterflies move into spots of sunlight that have penetrated to the forest floor and display by opening and closing their beautifully patterned wings in the bright spotlights. They also compete with each other for the best spot of sunlight.

Very little light filters through the canopy of leaves and branches in a rainforest to reach ground level, or close to the ground, and at those levels the yellow to green wavelengths predominate. A signal might be most easily seen if it is maximally bright. In the green to yellow lighting conditions of the lowest levels of the forest, yellow and green would be the brightest colors, but when an animal is signaling, these colors would not be very visible if the animal were sitting on a yellowish to greenish background. As John Endler explains, the best signal depends not only on its brightness but also on how well it contrasts with the background against which it must be seen (Endler, 1993). The green tree frogs that inhabit Australia's rainforests (15 different species of *Litoria*) are colored in shades of green on their backs and bright yellow or white on the undersurfaces of their bodies and on the limbs. These colors may provide camouflage in the lower levels of the rainforest. Colors close to, but not identical to, yellow and green are best for signaling in this locale because they are bright and can also be distinguished from the background. In this part of the forest, therefore, red and orange are the best colors for signaling, and they are the colors used in signals by the ground-walking Australian brush turkey *(Alectura lathami)*. This species, which lives in the rainforests and scrub lands of the east coast of Australia, has a brown to black plumage with bare bright red skin on the head and neck and a neck collar of orange-yellow, loosely hanging skin. During courtship and aggressive displays, the turkey enlarges its colored neck collar by inflating sacs in the neck region, and then flings about a pendulous part of the colored signaling apparatus as it utters calls designed to attract or repel. This impressive display is clearly visible in the light spectrum illuminating the forest floor.

In higher zones of the forest or in more open areas where trees have been felled, blue-gray illumination is predominant, and there blue or blue-green coloration is the brightest and red or orange again provides the best contrast. The blue-green and red combination of colors is exploited by the swift parrot *(Lathamus discolor)*, which inhabits open forest in southeastern Australia, and the eclectus parrot *(Eclectus roratorus)*, which lives in the upper levels of tropical rainforests and the adjacent eucalyptus woodlands of the most northerly tip of Queensland. Interestingly, the best signal colors are sported differently by the male and female eclectus parrots. The male is bright green with scarlet red flanks under the wings and a large orange-red beak, while the female is bright blue and scarlet with a black beak. The male has blue feathers in the wings but these are displayed only in flight.

Species that seek out small gaps in the canopy, where reddish light predominates, should signal with red, orange, or yellow for maximum brightness and use purple or blue for contrast. The male Australian rainbow lorikeet *(Trichoglossus haematodus)* uses this combination of colors in a clownish fashion: he has a blue to purplish head and underbelly, a bright red beak, and an orange and red breast, together with a green back and green upper wing surfaces. The green upper surface provides camouflage despite the conspicuous colors used for signaling because it disrupts the outline of the bird's shape; in particular, the green upper surface conceals the bird from aerial predators such as falcons or hawks. This species is a striking example of the outcome of evolutionary processes that have selected a balance between colors that will conceal and colors that can be used to signal.

Less colorful birds and other animals that inhabit the rainforest tend to rely on forms of signaling other than the visual, particularly over long distances. The piercing cries of the rhinoceros hornbill characterize the southeast Asian rainforest, as do the unmistakable calls of the gibbons. There is also the long, rather terrifying call of the male orangutan, which carries over considerable distances to advertise his presence. In densely wooded environments, sound is the best means of communication over distance because, in comparison with light, it travels with little impediment from trees and other vegetation. In forests, visual signals can be seen only at short distances, where they are not obstructed by trees. The male riflebird exploits both these modes of signaling simultaneously in his courtship display. The sounds made as each wing is opened carry extremely well over distance and advertise his presence widely. The ritualized visual display communicates in close quarters when the female has approached.

Under the Sea

Under the sea, too, light conditions are varied and constantly changing. As snorkelers know, shallow areas of the sea have ever-changing patterns of light of different wavelengths and intensity. The ability to change color, which many sea creatures posess, is a distinct advantage under the sea—it can be used to avoid being seen by predators. In this environment, too, social signaling commonly involves changing color. The cuttlefish not only changes color to camouflage itself against the background but also flashes color messages to other members of its own species, sometimes changing color only on the side of the body that its conspecifics (members of same species) will see as they swim past. The other side retains its camouflage pattern. This is an extreme case of directing the signal in precisely the desired direction, at the same time avoiding detection by other cuttlefish and predators.

As we have seen, sound signals also must be chosen according to the auditory environment. Sound can be heard over greater distances in certain conditions. For example, sound travels well in water and this is why whales use sound to signal over many miles. Sound is attenuated by vegetation and the surface of the sea floor. In general, high-pitched sounds are attenuated more than deep, low-pitched ones. So calls that need to advertise the presence of the sender over long distances should be both loud and deep, like the long-distance calls used by whales. The same principles apply to sounds transmitted in air, and thus the long call of the orangutan and the bellow of the elephant are both loud and low-pitched. High-pitched ultrasound is also used by some aquatic mammals (dolphins, for example) to navigate and find prey in murky or dark waters (Stebbins, 1983).

We have seen that olfaction is an important sense in fish communication. Although fish use visual and auditory signals as well, chemicals released into the water can be carried by currents over very long distances. They are ideal for long-distance communication underwater, although the direction of current flow limits the signal to downstream receivers. The males of some species of fish signal their presence to females by releasing chemicals into the water. When females detect the chemical signal, using their olfactory sense, they swim upstream toward the male.

Once the male comes into sight, visual signals play an additional role in beckoning the female.

In the Dark

Visual signals are ineffective in dark environments such as caves or burrows. Thus species living in these environments communicate by sounds or smells. Bats use ultrasound, sound of such high frequency (or pitch) that it is outside the hearing range of humans. They use ultrasound both to navigate in the dark and to communicate with each other. Cave-dwelling oilbirds *(Steatornis caripensis)* and swiftlets (*Collocalia* sp.) also use ultrasound to navigate and communicate when they are inside the dark caves where they nest, although they use vision outside the cave.

In their underground burrows, moles and rats may communicate by sound. In fact, vision is so unimportant in this environment that one burrowing species, the mole-rat *(Spadix ehrenbergi),* has effectively no eyes. Through the course of time and the process of evolution, the eyes have become minuscule and the skin and fur have grown over them. Even the external ears are not detectible. The mole-rat has become a cylindrical-shaped animal with short legs and tail, the perfect design for moving along tunnels only just big enough for it. These animals communicate with each other by tapping their snouts on the walls of the burrow. The vibrations are seismic signals that can be detected by a mole-rat in another tunnel of the burrow even if it is quite a distance away. At closer quarters the mole-rats communicate by vocalizing rather than tapping, and they also use odors. A study by Uri Shanas and Joseph Terkel (1997) has demonstrated that mole-rats release an odorous secretion from a gland in the orbit of the eye when they groom themselves. The secretion runs down a duct and out through the nostrils. By grooming, the mole-rats spread the secretion over their bodies and it serves to decrease aggression between males. In effect, the odor signals, "Don't fight".

Communication by seismic vibrations is also common in nocturnal desert rodents. North American kangaroo rats *(Dipodomys)* and African gerbils (types of rat, including *Gerbillus* and other genera) strike their feet against the ground to produce drumrolls that characterize the individual's species. Jan Randall has shown that kangaroo rats have individual signature rhythms that communicate the animal's identity and lower the risk of disputes over territory among neighbors (Randall, 1997).

The white-lipped frog of Puerto Rico *(Leptodactylus albilabris)* is believed to have the greatest sensitivity to seismic stimuli of all known species. The male embeds himself in mud and produces advertisement chirps or aggressive chuckles; as he expands his vocal sac in order to make the chirp call, the sac strikes against the muddy substrate to produce a seismic "thump" that is detected by other members of his species in the vicinity. These signals may be used by nearby males to synchronize their calls when singing in chorus and to maintain separation between individual frogs (Narins, 1990).

In addition to sound and odor, animals may use electrical signals to communicate in the dark. Electrical signals are an effective mode of communication in the murky waters of streams, and they are used by the electric fish of South and Central America to navigate and communicate with each other. We have already seen the way in which electric fish communicate their sex and dominance.

MEASURING COMMUNICATION IN ANIMALS

It is not always simple to prove that communication has occurred in animals, and it is even more difficult to decipher exactly what has been communicated. First we need to know a lot about the behavior of the species we are investigating, and then we have to use certain techniques to determine whether a signal has been sent and received. We may detect that communication has occurred by observing behavior to see whether a particular activity performed by one animal consistently leads to a change in the behavior of another animal, or animals. This requires very careful scrutiny, and observations must be repeated many times. Once the initial observations have been made, they can be followed up by experiments designed to determine the exact nature of the communication. There are several clever ways of proceeding.

Audio Playback Experiments

One of the main ways to study communication in animals is to record the signal of interest and then play it back to the animals and see whether they respond in a predictable way. For example, many songbirds sing to advertise their territory. These territorial songs can be recorded on audiotape and then played back over and over again through a loudspeaker placed in an unoccupied territory. If males of the species stay out of the area where the loudspeaker is located, it may be concluded that the song is indeed a territorial vocalization. Of course, it is not as simple as this because we need to have an experimental control. We need to know how rapidly males would move into an unoccupied territory without a loudspeaker broadcasting the song.

Experiments of this type have demonstrated that the European great tit *(Parus major)* produces a specific territorial song. John Krebs removed pairs of great tits from their territories in a forest and then placed a loudspeaker broadcasting the song of the great tits in some of the territories and left other territories empty. He found that the territories without loudspeakers were reoccupied far sooner than those with the loudspeakers broadcasting the song (Krebs, 1977). This experiment shows that the song does advertise that a given territory is taken and warns other males of the species to stay out of it. But the fact that the territories with loudspeakers were eventually occupied suggests that continued maintenance of a territory requires more than simply singing in one spot. It might require that the birds move around in the territory and use visual displays to accompany the song.

Even though this particular experimental procedure can demonstrate that a song advertises territory and signals to other males to keep out, it is important to go a step further to see how specific the song has to be in order to communicate this signal effectively. This can be done by playing another song, or a modified version of the original song, through loudspeakers placed in unoccupied territories to see whether these sounds also inhibit males of the species from moving in to occupy the territory. If these songs do not keep males out of the territory, or if they are clearly less effective in doing so than the original song, we can conclude that there is some specificity in the song. If, however, sounds other than the song also keep males out of the area around the speaker, there is no such specificity, and we would be unable to conclude that the song itself was communicating territory ownership.

With this technique, it is possible to determine exactly what aspects of the song convey the important information about territory ownership. This can be done by modifying the recorded song in various ways. Parts of it might be left out, or the song could be played backward. Alternatively, the order or sequence of the syllables (parts of the song; see Figure 2.1) may be changed. There are many ways of modifying the song. The effect of playing back the modified song can then be compared with the effect of playing back the unmodified song. In this manner, it is possible to single out the essential aspects of the song that warn other males to keep out of the territory.

John Krebs followed up his first experiment by playing back modified songs. In fact, he noticed that the great tit has a repertoire of several songs. One male may sing up to eight different types of song. Since individual birds vary in how many song types they sing, he was interested to see whether the size of the repertoire would alter the signal, making it more or less effective. To do this, he located loudspeakers that played back only one song in some unoccupied territories and speakers that played back a repertoire of up to eight songs in other territories. The territories in which the larger repertoire was broadcast were reoccupied after a much longer delay than those in which the smaller repertoire was played (Krebs, Ashcroft, and Webber, 1978). This experiment demonstrated that singing more song types together in a repertoire is a more effective signal than singing only one song type. Hence males with more elaborate songs can maintain their territory more effectively than those with less elaborate ones.

The fact that variations of a song produce different results raises another issue about the design of playback experiments, as Donald Kroodsma (1990) first realized. It is important to select many different songs to play back. In some of the early playback experiments, only one song, or very few songs, were played through the loudspeaker, and this lack of of variety could have seriously limited the results. In fact, many avian species learn to recognize the territorial songs of other members of their species holding territories next to their own and respond differently to the territorial calls of their immediate neighbors than to those of birds from more distant territories.

Emma Brindleym has investigated the responses of European robins *(Erithacus rubecula)* to the songs of neighbors and strangers (1991). Despite the large and complex song repertoire of European robins, they were able to discriminate between the songs of neighbors and strangers. When they heard a tape recording of a stranger, they began to sing sooner, sang more songs, and overlapped their songs with the playback more often than they did on hearing a neighbor's song. As Brindley suggests, the overlapping of song may be an aggressive response. However, this difference in responding to neighbor versus stranger occurred only when the neighbor's song was played by a loudspeaker placed at the boundary between that neighbor's territory and the territory of the bird being tested. If the same neighbor's song was played at another boundary, one separating the territory of the test subject from another neighbor, it was treated as the call of a stranger. Not only does this result demonstrate that the robins associate locality with familiar songs, but it also shows that the choice of songs used in playback experiments is highly important.

The playback technique can be used to study the territorial vocalizations of other species, and it can also be used to study other kinds of auditory signals. For example, Jan Randall determined the meaning of the drumrolls made by kangaroo rats by playing back foot-drumming recordings of three different species to wild populations of each species. Two of the species (*Dipodomys spectabilis* and *D. ingens*) responded to hearing the playback by drumming and the other species (*D. desertii*) approached the loudspeaker. These responses to the playback are typical of each species: *D. desertii* chases intruders away and only rarely drums the feet, whereas the other two species engage in drumming exchanges. Thus each species responds to hearing the playback of sounds used in communicating about identity and territory in ways typical of the species (Randall, 1994, 1997).

The playback technique can be used to investigate other forms of communication, not just those about territory. For example, playing the songs of male canaries to female canaries stimulates them to build nests. Alternatively, playing the warning call of a species stimulates appropriate evasive action. Christopher Evans and Peter Marler (1993) found that the alarm call of cockerels differs depending on whether they see a predator on the ground or in the air. When they see a hawk, or even a hawklike image, moving overhead, they emit a long screech (Figure 2.3A), which is entirely different from the call given when they see a predator on the ground, such as a dog or raccoon (Figure 2.3B). The latter is a repeated pattern of short pulses ending with a little flourish. It should be noted that the warning call signaling the presence of an aerial predator is a thin high-pitched sound, as is the warning call of the galah and many other species. The source of such calls is difficult to locate, and hence the caller is less likely to be detected by an aerial predator.

Having recorded these two calls made by the cockerels, Evans and Marler used the playback technique to assess whether the calls signaled anything specific to other chickens. They tested each chicken individually

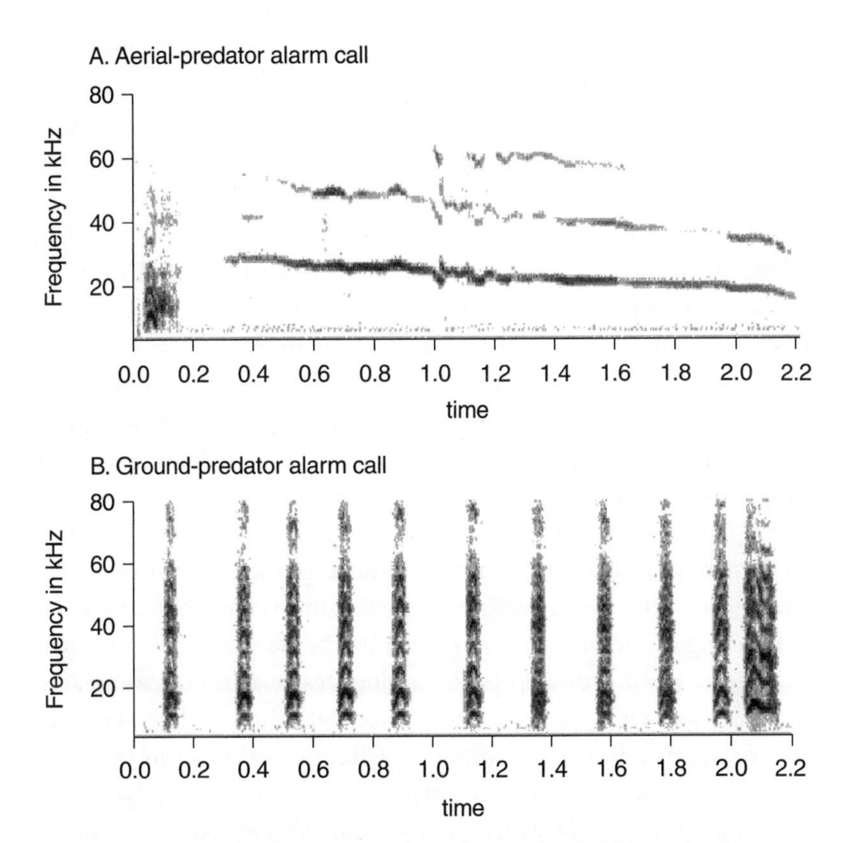

FIGURE 2.3 Sound spectrograms of the alarm calls of chickens (*Gallus gallus*). Chickens produce different calls to signal the presence of predators on the ground and predators in the air. A: The alarm call given when an aerial predator (hawk) is seen flying overhead. B: The alarm call given when a ground predator (dog or raccoon) is seen. (Adaptations of recordings made by C. S. Evans and P. Marler; sound spectrograms courtesy of C. S. Evans.)

in a cage in the laboratory, where it could not see any predators and was not exposed to any other changing visual stimulus that might cause it to vocalize. They then played the two kinds of alarm signals through a loudspeaker. When the aerial alarm call was played back, a chicken hearing it would crouch and look up as if trying to catch sight of the predator in the air. When the ground-predator alarm call was played, the chicken hearing it would run for cover or strut while calling in a way that might drive the predator away. Thus the two alarm calls have specificity and signal to the receiver to take appropriate measures to avoid being caught.

Vervet monkeys *(Cercopithecus aethiops)* also produce different vocalizations for different predators. Males make a deep barking call for a leopard and females make short, high-pitched chirps in the same circumstance. A chutter-like call is made for a snake and a single cough-like call for an eagle. Dorothy Cheney and Robert Seyfarth carried out playback experiments at a field site in Africa. They found that the vervet monkeys took the appropriate evasive action for the predator indicated by the call (Cheney and Seyfarth, 1990; Seyfarth, Cheney, and Marler, 1980). When the leopard call was played, they dashed to the nearest tree and climbed it. On hearing the snake call, the monkeys stood up on their hind limbs and peered into the grass. When the eagle call was played, they looked up and took cover, behavior not dissimilar to the evasive action taken by the chicken on hearing its own species' aerial alarm call.

Living in the same territory as the vervet monkeys are superb starlings, and these birds also make different calls for eagles and terrestrial predators. Cheney and Seyfarth discovered that the vervet monkeys knew the meaning of the predator calls made by the starlings as well as their own species-specific calls. When the starling's eagle alarm call was played over a loudspeaker, the monkeys looked up; when the starling's ground-predator alarm call was played, most of the monkeys ran to the trees. No response was given when the starling's song was played back, and that was a control for the experiment because the song does not indicate the presence of any predator. Apparently the monkeys have learned to interpret the alarm signals of the starling. In this way different species living in the same area may make use of each other's communication signals.

Video Playback Experiments

Auditory signals have so far been the main focus of attention in the playback technique, but recent advances in video imaging have made the technique applicable to visual signals. It is now possible to make video recordings easily and, through digital manipulation, to change the recorded image for playback. Thus the behavior thought to be a visual signal can be recorded on video and played back to another member of the species in a controlled setting to see whether it elicits a reliable response in the receiver. Then, just as for the playback of altered songs, components of the video image can be eliminated or modified to determine exactly what aspects of the image are essential for the signal.

Of course, this technique is successful only if the species being studied pays attention to video images, which are two-dimensional and flicker. Humans cannot see flicker when it is very fast but some animals can see flicker at frequencies that we cannot detect. Fluorescent lights flicker on and off at so fast a rate that we perceive the light as being on continuously. But some birds may be able to see the flicker, and it would appear to them as a strobe light does to us. Video playback may also be seen to flicker by some species, and this would make it far from suitable for testing those species.

Nevertheless, Evans and Marler found that chickens do attend to video images in experiments where they used the video image of a chicken as a companion to a rooster stimulated to give alarm calls. First, they found that a rooster is much more likely to emit an alarm call on seeing an aerial predator when another live chicken is present in an adjoining cage. Next, they were able to show that replay of a video image of a chicken, with an accompanying soundtrack, would have the same effect as a live chicken of enhancing alarm calling by the rooster. Then they played a video image of another species, a bobwhite quail, and found it to be less effective than the video image of the rooster's own species (Evans and Marler, 1991).

Video imaging has also been used to study the visual signals made by male lizards in courtship and aggressive encounters. Joseph Macedonia, Christopher Evans, and Jonathan Losos used video playback to investigate head-bobbing and pulsing of the dewlap, the skin under the chin that can be extended and contracted, performed by two species of *Anolis* when males encounter each other. These are aggressive visual displays. The researchers found that video images of head-bobbing and dewlap displays elicited similar displays by the live lizards watching them, and that seeing a video of a member of the same species was more effective in eliciting the aggressive display than seeing a video of another species. This illustrates at least some degree of species specificity in visual signals of aggression (Macedonia, Evans, and Losos, 1994).

Video recording and playback can also be used to study communication by facial expression in animals. The image can be varied by changing the eyes, nose, mouth, or other features, either by distorting its contribution to the total facial expression or by eliminating each feature in turn from the expression. In this manner, the relative importance of the various features in any particular facial expression can be determined. For example, orangutans, and many other primates, perform one kind of play-threat display by opening the mouth, puffing up the lips, showing the teeth, and raising the eyebrows (Figure 2.4). It is not known which (if not all) of these features signals to another orangutan, but now it would be possible to find out by using video playback of manipulated images.

Playing video images to animals to study signaling has the advantage of being able to repeat exactly the same sequence of videotaped recordings as many times as the experimenter wishes. Exactly the same sequence can be presented to different animals, or again and again to the same animal, to test the reliability of the signal.

CONCLUSION

When communicating, different species use their different sensory systems to varying extents, depending on where they live and the most effective way to send a signal. We may say that communication is entirely context-dependent, meaning that what is communicated and when communication occurs depends on the environmental context surrounding the animal.

As observers of animals, we must first establish whether communication has actually occurred by determining whether a signal sent by one animal changes the behavior of the animal receiving it. Then we must determine exactly what has been signaled. Most communication between animals must depend on the use of specific signals that are not ambiguous, but there are examples of the same signal being used in entirely different contexts and causing quite different responses by the receiver. These signals would seem to be less specific, unless the same signal means something different when it is given in a different context, or the signals are, in fact, different in subtle ways that have eluded us. By using the playback technique and modifying the calls played back, we should eventually be able to distinguish between these alternatives.

FIGURE 2.4 Open-mouth play-threat display of an orangutan. Note the puffed area around the mouth, the bared teeth, and the raised eyebrows. This signal was directed toward another orangutan and it was followed by a play attack on that orangutan. (Photograph by G. Kaplan.)

Chapter Three

IS SIGNALING INTENTIONAL OR UNINTENTIONAL?

We humans do not always communicate verbally. Sometimes we communicate vocally, using sounds that are not words, and sometimes we communicate by touching another person. And almost always, whether we are speaking or not, we create facial expressions. These forms of communication are all nonverbal. Nonverbal communication may be intentional or unintentional. But a large amount of the communicating that we do by nonverbal utterances or facial expressions is unintentional, signaling something about our internal (emotional) state.

The fact that there are many different vocalizations that humans emit without any intention of communicating raises the possibility that all or most vocalizations made by animals are also unintentional utterances. Those who believe that signaling by animals is unintentional make an absolute division between animals and humans, reserving intentional communication for humans alone. They believe that animals simply emit vocalizations, and other signals, unthinkingly. They assume that the vocalizations and other signals produced by animals are involuntary, that they cannot be controlled consciously and are simply generated as automatic expressions of the animals' internal states.

Those who hold this view say that a chick peeps when it is cold and twitters when it finds a warm place as if it were a little machine, not because it actually wants to communicate that it feels distress or pleasure. In fact, the divide between animals and humans is widened even further by those who assume that animals may not be aware of feeling distress, pleasure, or any other internal state. The animal is seen as a robot without feelings or the ability to think, let alone communicate intentionally.

This view of animals has an equally mechanistic explanation of the animal that receives and responds to the signal. The hen might respond to the chick's distress calls by leading it to food but she does so without thinking, not knowing what she has heard or even that she is responding. The interaction between the chick and the hen is interpreted as simply one little robot making a sound that causes a slightly larger robot to change its behavior.

To refute this attitude about animals we would need to consider whether animals are capable of thinking for themselves and whether they can actually feel things that they communicate to others. It is not our intention here to explore the broad topic of thinking and awareness in animals (see *Minds of Their Own* by Lesley Rogers, 1997b), but we will consider those aspects of communication that tell us something about whether or not animals communicate intentionally.

The fact that a signal is sent by one animal and that it leads to a change in the behavior of another animal is not, in itself, evidence that the sender intended to communicate or that the receiver intended to respond. It is very difficult to prove that the animal sending the signal intends to communicate, but there is new evidence showing that animals communicate specific information in specific circumstances. These findings work against the notion that animals emit signals simply as a reflection of their emotional state and in an uncontrolled and random way. Emotions certainly play a role in signaling by animals, as they do in humans, but communication in animals is not simply an automatic expression of the emotions, as we discuss below (see also Marler and Evans, 1996).

THE EFFECT OF HAVING AN AUDIENCE

If an animal communicates its internal, emotional state unintentionally, it might be expected to signal in exactly the same way whether it is alone or in the company of other animals; if it communicates intentionally, we might expect it to confine its signaling to occasions when it has an audience.

Warning calls are a special case for considering the presence or absence of an audience. Let us consider the case of a chicken *(Gallus gallus)* that emits a warning call when it catches sight of a hawk flying overhead. In Chapter 2, we saw that cockerels make one warning call for a predator seen flying overhead and a different call for one approaching on the ground (the aerial-predator versus the ground-predator alarm call: Figure 2.3). Thus a cockerel makes a specific screeching call when he sees the aerial predator but, by doing so, he draws attention to himself and increases the chance that he will be taken by the predator. By issuing the warning call, the individual may save the group but risks his own life.

There has been much debate about whether this is a genuinely altruistic act on the part of the caller or whether it is not particularly altruistic because the individual shares some of his genes with other members of the flock and, therefore, by calling he increases the chance of survival of those genes. We will consider whether the individual issuing the warning call does so intentionally or simply emits the vocalization when his internal state is changed by seeing the predator.

Obviously, the bird that sees the predator feels fear. The call issued could simply be an expression of that internal state of fear—an automatic, unintentional signal of the chicken's emotional state. The first piece of evidence to indicate that this is not the case is the fact that the chicken gives different calls for aerial and ground predators although both predators induce a state of fear. It could, however, be suggested that an aerial predator elicits more fear than a ground predator (or vice versa) and that the different calls are merely a reflection of the amount of fear that the individual feels. Switching from one call to another completely different call as the internal state of fear increases does appear to occur in some species, as we will see later in this chapter. In other species, however, increasing states of arousal (fear) are accompanied by the same call being issued more often or more loudly.

The second piece of evidence against the idea that the two different alarm calls are emitted unintentionally is the fact that the presence or absence of an audience influences whether calling occurs or not. If the cockerel happened to be alone when he saw the predator, there would be no advantage in making a warning call. In fact, to issue a warning in this circumstance would be nothing less than disadvantageous. But if the cockerel cannot control his vocalizations and merely emits the warning call as a reflection of his particular internal state at the time, he will call irrespective of whether he is alone or in the presence of other members of his species. In contrast, if the cockerel can control his vocalizations and raises the alarm only when he has the intention of warning other chickens, he should not call when alone.

The latter is the case, as Peter Marler and his colleagues have shown.

They measured the alarm calls made by a cockerel in a cage with a video monitor placed overhead. When an image of a hawk in flight, or an approximation of one, was presented on the video monitor the cockerel made aerial-predator alarm calls, but only when there was a male or female of his own species in a nearby cage (Karakashian, Gyger, and Marler, 1988). He rarely uttered an alarm call when the same cage was empty or when it contained a bird from a related but different species (they used a bobwhite quail). An audience of the cockerel's own species had to be present for normal levels of alarm calling to occur. As we saw in Chapter 2, Marler and Evans were later able to show that a video image of a hen of the same species, instead of a live hen, would also serve as an audience.

The need for an audience before aerial-predator alarm calling will occur shows that the cockerel is not a simple robot emitting alarm calls when he is triggered by the appropriate stimulus (the predator). The social environment is taken into account before calling occurs. It could be said that the cockerel does not call when alone because he does not become sufficiently aroused by seeing the hawk unless another chicken is present. This could mean that he does not call intentionally but requires two conditions to be met before he will call automatically. Such an explanation will suit those who wish to make an absolute distinction between the communication systems of animals and humans. However, it does not seem to be correct, because the cockerels showed the same amount of looking overhead, crouching, immobility, scuttling away, and sleeking down of their feathers with and without an audience. Apart from the absence of alarm calling, the cockerels when alone reacted to the hawk to the same extent, irrespective of the presence or absence of the hen (Marler and Evans, 1996). This shows that they were, in fact, just as afraid when alone as when they had an audience. We may therefore conclude that the cockerel actively suppresses alarm calling when alone. We think the most likely interpretation of these findings is that the cockerel makes the warning call only when there is a reason for doing so and that, when he does call, he does so with the intention of warning other members of his own species, most likely kin. More experiments will be necessary before we can be sure of this explanation.

As we saw in Chapter 2, vervet monkeys give different alarm calls for different predators. The presence of an audience also determines whether they give alarm calls. Solitary vervet monkeys have been observed to escape from an approaching leopard in total silence (Cheney and Seyfarth, 1990). Apparently, the absence of other vervet monkeys negated the need to call and alarm calling was suppressed, just as with the cockerel.

Cheney and Seyfarth (1985, 1990) also conducted experiments on captive vervet monkeys, demonstrating that adult females give more alarm calls when their offspring are present. Despite this effect of audience, however, Cheney and Seyfarth concluded that vervet monkeys do not know anything about the mind-state of their audience—whether their audience knows or does not know that a predator is nearby. The researchers based this conclusion on the fact that the monkeys go on making alarm calls long after every other monkey in the group has seen the predator; and in the presence of their offspring, mothers call no more often, or differently, for predators that offer great threat to their offspring than for those that pose a lesser threat.

Although these observations suggest that the signaler does not differentiate its own vulnerability from that of its audience, more information is needed before a firm conclusion can be reached about the ability of the monkeys to modulate calling according to the state of knowledge of group members. But it is clear that they can vary alarm calling according to the presence or absence of an audience. They do not signal impulsively and involuntarily but decide whether to signal or not and what signal they will use in a given context.

The presence of an audience also increases the calls that cockerels produce in the presence of food (Evans and Marler, 1994). There is a typical food call, consisting of repeated pulses of sound, that the cockerel produces when he sees food, or another stimulus that he associates with food, and this call attracts hens. The hens run to the male and the male drops the food, allowing the hens to eat it. This behavior is often followed by courting and mating. In fact, food calling is enhanced by the presence of a hen.

Evans and Marler (1994) were able to show that the enhancement of calling in the presence of the hen is not just a general effect on the motivation (or arousal) of the cockerel to feed, because the presence of the hen increases the food calling but not the pecking at the food. In other words, the presence of the hen had a specific effect on signaling about food but did not have any effect on behavior not used for signaling. It would seem, therefore, that the cockerel signals with the intention of alerting the hen to the presence of food and does not simply emit calls automatically when he sees food.

In addition, Evans (1997) has now shown that the hen looks for food on the ground when she hears the food call. Evans observed the behavior of a hen when she was played the food call through a loudspeaker. On hearing the call, she put her head close to the floor and walked around as if looking for food, even though there were no grains of food on the bottom of the cage in which she was tested. Thus her searching for food was triggered by hearing the food call specifically and not by her having caught sight of any grains of food. The receiver of the signal responded in a specific manner.

It is most important to note that either a male or female conspecific is an effective audience for a cockerel to signal the presence of an aerial predator, whereas only a female is an effective audience for the food call. This demonstrates even more specificity of the situation in which the cockerel will produce calls, and this specificity is also necessary for the survival of the species. Both males and females can benefit by being alerted to the presence of an aerial predator, but food calling is used by males to attract females as a prelude to courtship. In other words, the effect of the audience is not simply a matter of its presence or absence; the audience has specific relevance to the particular social context and, it would seem, to the intent of the signaler.

In the examples given so far, calling is enhanced by the presence of an audience, but this is not the case for all types of calling. The presence of an audience has no effect on the amount of calling the cockerel gives when he sees a predator on the ground. Marler and Evans (1996) reasoned that this is so because the ground-predator alarm call is used not only to alert other members of the species to the presence of a predator but also to confront the predator itself, and to try to drive it away. This contrasts with the aerial-predator alarm call, which is associated with behavior that would hide the bird from the predator. The aerial-predator alarm call is a thin sound that fades in and out, making its source very difficult to locate, whereas the ground-predator alarm call is conspicuous, abrupt, repeated, and easy to locate. It appears to be designed to capture the ground predator's attention and is accompanied by behavior that might make the predator decide to look elsewhere for a meal.

Drawing attention to itself would appear to be the best strategy for a bird to adopt when confronted by a predator on the ground. Chickens can fly far enough to get away and they can run fast too, but the strutting and calling display directed at the predator might be a less energetically costly way of signaling to the predator that the bird could escape if approached. This maneuver may resemble the stotting of Thomson's gazelles when faced with predators. The stotting signals how fit the gazelles are and thus how they could escape, as we saw in Chapter 1.

Fleeing would be the only alternative available to the chicken approached by a predator on the ground. This strategy might follow the strutting and alarm calling at a moment when the bird's calling and strutting activity has put the predator off guard. Whatever the reason for the bird's strutting around when confronted by a predator on the ground, there is as much reason to strut around and make the ground-predator alarm call when an audience is present as when it is not. This might explain why having an audience has no effect on ground-predator alarm calling.

The difference of the audience effect on aerial-predator alarm calling versus ground-predator alarm calling illustrates that calling is very specific to the context in which it occurs, and that there is no single, simple set of rules that the bird follows to control its vocalizations. Although this does not prove beyond all doubt that birds communicate intentionally, it certainly indicates that they may do so. Also, there is further evidence to suggest that chickens can control when and what they communicate, and this finding concerns the use of calls to deceive another individual, as we will discuss next.

DECEPTION

The use of signals to deceive another is perhaps the most sophisticated form of signaling. In the most developed form of deception, the deceiver may know the usual context of the signal and then use it in an unusual context with the intent of deceiving another animal. It appears that animals sometimes communicate deceptively, but it is very difficult to prove beyond doubt that they have done so intentionally. Nevertheless, there is some evidence indicating that animals do engage in deception with intent, as outlined by Rogers (1997b).

Marcel Gyger and Peter Marler (1988) have observed that cockerels sometimes make food calls when no food is present. They appear to do this only when the hen is far enough away that she cannot see whether food is actually present where the cockerel is located. On hearing the call, the hen approaches the cockerel, presumably to search for food in his vicinity. Thus, by issuing the food call when no food is present, the cockerel can deceive the hen into approaching provided she is so far away that she cannot see that he is cheating. According to Gyger and Marler, the cockerel does not use food calls deceptively when the hen is close enough to see that he is signaling deceptively.

These are interesting observations, but more experiments need to be carried out to decide whether the calls are indeed being used deceptively. It is possible, for instance, that the cockerel emits food calls to attract the hen only when she is farther away because he is more motivated to obtain her company when she is at a greater distance, rather than because he has figured out that he can deceive her only when she is farther away. Nevertheless, studies like Gyger and Marler's are important and interesting approaches that attempt to unravel the difficult problem of intentionality in animal communication, and they lay the groundwork for more research in the area (see the 1997 paper by Christopher Evans for more discussion).

Other examples of deception have been reported by ethologists studying the behavior of animals in their natural environment. We will not list them all here, but we do draw attention to one form of deception that has been observed in many different species: issuing a warning call or behaving as if a predator were nearby when there is no evidence that that is the case. We refer to this behavior as "crying wolf," after the story of the boy who cried "Wolf!" too often and so was ignored when he really needed help—an excellent example of habituation in the receivers.

Predator-warning behavior appears to be used in many species to distract the attention of the receiver who is eating a favored food; after diverting the receiver's attention, the deceiver moves in to grab the food for itself. The Arctic fox has been observed to use warning calls in this manner, and so have domestic dogs and certain species of birds (described in detail in Rogers' *Minds of Their Own,* 1997b). In *The Thinking Ape,* Richard Byrne (1995) describes many observations of deception in primates. One incident involved the pretense that a predator was nearby: a baboon being chased by another baboon was observed to stop and look around as if there were a lion or other predator in the near distance, and when it did so, its pursuer stopped and looked around too, giving the pursued baboon time to escape. Seeing no evidence of a predator in the area, Byrne interpreted this behavior as deception. The pursued individual had manipulated the pursuer by signaling incorrect information.

Deception is perhaps the most complex form of communication. It can occur only when a communication system is firmly in place and usually functions in a consistent and reliable (referred to as "honest") fashion. Individuals who are detected signaling dishonestly are punished or their signals are ignored. Deception is a risky form of communication. Its existence suggests the intentionality of communication.

Mimicry is a form of deception. Wolfgang Wickler (1968) has described the way in which certain nonpoisonous butterflies mimic the appearance of poisonous ones to gain protection from avian predators that have learned to avoid the poisonous butterflies. In this case, the deception is definitely unintentional. Other forms of mimicry may, however, be intentional. We do not yet know for certain, but some forms of vocal mimicry in birds may be used to deceive predators and such behavior may well be intentional. By mimicking the vocalizations of their predators, some avian species may signal that the territory is occupied by another member of the predator's species and so prevent the predator from moving in. There is some evidence that this occurs, but much more research is needed before we can say anything conclusive about the behavior. We will discuss mimicry in more detail later.

ALARM CALLS THAT REFER TO PREDATORS

We have discussed the different calls produced by chickens and vervet monkeys to warn other members of their species of specific classes of predators. These signals are called referential signals because they appear to be analogous to human words used to refer to animals, objects, or events. A number of other species also use different calls to refer to different types of predators (see Macedonia and Evans, 1993).

In a refinement of this behavior, prairie dogs *(Cynomys gunnisoni)* can actually signal the details of a predator in their alarm calls. Con Slobodchikoff and his colleagues (1991) recorded the alarm calls that prairie dogs made as humans approached at a walking pace. Since this species has been preyed on by humans for more than a hundred years, the experiment was relatively natural, or at least relevant to the species. The human "predators" wore different clothes in different tests, white laboratory coats or colored shirts, and different people were involved. By recording the calls made by the prairie dogs and then analyzing different detailed features of the calls, the researchers were able to show that the prairie dogs may be able to distinguish one human being from another and that they may incorporate information about the physical features of individual predators into their alarm calls. This is an interesting result, but it needs to be supported by tests showing that the prairie dogs actually use this information when they hear the different signals. If so, these animals not only are perceiving much more detail than we might have thought but also are signaling this information to each other. By using playback experiments it should be possible to see whether the prairie dogs actually use the detailed information encoded in the alarm signals.

Ring-tailed lemurs *(Lemur catta)* produce differentiated alarm calls to alert their group members to an aerial or a ground predator, but they also have a general call, a relatively soft "glup" sound, that they produce when they first catch sight of any predator or, indeed, when they perceive any startling visual or auditory stimulus. This seems to be a general alert signal to the group. If an aerial predator has been detected, they follow the "glup" with loud calls, first rasps and then shrieks when the predator is within attack range. If the predator is a carnivore (ground predator), the "glup" is followed by "clicks" and "yaps." Thus the lemurs signal information about aerial versus ground predators and also about the proximity of the predator.

The ground squirrels of California *(Spermophilus beecheyi)* produce "chatter" calls when they see a predator on the ground and "whistles" when an eagle or hawk flies overhead, but their calls are not as specific as the alarm calls of chickens or the eagle and snake alarm calls of vervet monkeys. Sometimes they chatter when they see a hawk in the distance or whistle when they are being chased by a carnivore. These apparent errors in reference to specific predators may, in fact, be conveying more detail about the situation in general. These calls may communicate the urgency of the situation and so convey information about potential versus imminent danger rather than simply indicating that the predator is a hawk or a dog. Thus while some species have specific calls to refer to different predators, just as we use different words for different animals, others may signal the urgency of the situation instead.

THE ROLE OF EMOTION

We have presented evidence showing that animals do not simply vocalize in an uncontrolled manner as a way of expressing their emotions. Nevertheless, their emotional state does affect their signaling, just as emotions affect speech and other forms of communication in humans. The variation in calls made by ground squirrels may, as mentioned above, indicate the urgency of the situation, conveying the proximity of the predator and perhaps also the behavior of the predator. The emotional state of the signaler may be the factor that determines these variations in signaling. When the squirrel is very afraid it may whistle, and when it is only mildly afraid it may chatter. In general, hawks may be a greater threat than ground predators and thus more likely to elicit high levels of fear and therefore whistle calls, but when a hawk is far away instead of nearby it elicits only a chatter. By contrast, being chased by a carnivore is a highly fear-inducing situation and the squirrels whistle when they are so threatened.

Similar systems of calling have been reported for other species: for example, the black-winged stilt *(Himantopus himantopus)*, a wading bird, has two types of alarm signal, depending on the distance of the predator from the bird's location. Again, increasing fear may lead to a switch from one call type to another. Similarly, as mentioned above, ring-tailed lemurs give rasping calls when an aerial predator is far away and shrieks when it is closer.

Other species vary the rate of calling as the risk of being caught by a predator increases, as has been documented for yellow-bellied marmots *(Marmota flaviventris)*. At field sites in Colorado and Utah, Daniel Blumstein and Kenneth Armitage (1997) studied the alarm calls the marmots gave on the approach of a trained dog, a model badger, a radio-controlled badger, and a walking person. The marmots made three different alarm calls, but the calls did not appear to be specifically related to any particular type of predator, possibly because all the "predators" used in the study were artificial ones. But the marmots' rate of calling increased as the predator came closer. Calling rate is therefore an indication of the level of fear. By playing back one of these calls at various rates, the researchers were able to show that the rate of calling did, in fact, signal the degree of risk to other marmots. Thus the receiver could interpret the meaning of the call from the calling rate.

Emotional state can be influenced by hormones and this can be reflected in signaling. The amount of aerial-predator alarm calling by cockerels is influenced by the level of testosterone circulating in the bloodstream, possibly because the hormone changes the bird's emotional state and the way it attends to the predator.

In humans we can tell the level of emotion by the intensity and quality of the voice. This may also be the case in animals, but the topic has not been studied to any great extent. We do, however, know that calling is more frequent and louder when animals are more aroused. The more distressed a young chick feels, the more often it peeps and the louder it peeps. A similar pattern of responses accompanies increased distress in a wide range of species, including humans. Emotional aspects of signaling may also be conveyed by other methods of communication accompanying vocalizations. Humans signal their emotional state when speaking by body posture and facial expression. A twitch of muscles in the face or a wringing of the hands can be more informative than the actual words being spoken. Animals too accompany their vocalizations with other signals that may indicate emotional state. Cockatoos, for example, raise their crests while they vocalize when they are alarmed.

In some cases, the behavior accompanying a particular vocalization is quite obviously another direct response to the stimulus that elicited the vocalization. Ground squirrels scurry into their burrows while giving their whistle calls. Vervet monkeys stand on their hind limbs and look down into the grass at the same time as they give the snake alarm call, look up and take to cover when giving the eagle call, and scamper up a tree when giving the leopard call. These characteristic actions accompanying each alarm call add to the power of its meaning, and the speed or vigor with which they are performed may indicate the amount of fear that the signaler is feeling, although this has not yet been studied.

More subtle behavioral changes may also accompany vocalizations. As Eckard Hess (1965) showed three decades ago, in humans, the size of the pupils in the eyes varies with emotional state and attitude. Also, humans assess the pupil size of other people with whom they are interacting, although they do so quite unconsciously. A greeting accompanied by dilation of the pupils is rated positively, whereas one with constriction of the pupils is rated negatively and viewed with distrust.

Pupil size may be an important aspect of communication in animals also. We know that it varies with the state of arousal or emotion. Some years ago Richard Gregory and Prue Hopkins (1974) reported that the pupil size of a parrot constricted whenever she produced learned words and also while she was listening to familiar words. There has been no research to test whether other parrots respond to the changes in pupil size, but it is potentially possible that they do.

Other emotional responses such as the erection of hair or feathers may also accompany vocalizations and signal emotional content. We have already mentioned raising of the crest in cockatoos. Most readers will be familiar with the way dogs raise the hair on their backs during aggressive encounters. Unfortunately, most studies of communication in animals focus on only one aspect of signaling and ignore the complete picture, so there is little detailed information on these added aspects of signaling.

THE ROLE OF COGNITION

Despite the contribution of emotions to signaling in animals, we must emphasize that animal signaling is not purely the expression of emotions, which are controlled at lower levels of brain function. Some signaling involves more complex cognitive processes in addition to those used to express emotions. By cognitive processes we mean higher levels of brain function, those that involve decision making, memory, and assessment of the environment. It is possible that some signals given by some species are purely emotional, emitted without cognition. But it is likely that most vocalizations involve both emotional and cognitive processes, although the emotional contribution may be greater in some signals and the cognitive contribution higher in others. The balance between emotion and cognition will vary with the function of the signal and the context in which it is given. This is likely to be as true for the vocalizations of animals as it is for those of humans.

The emotional content of human speech holds our interest and adds to the meaning of the communication, as is clearly demonstrated by the contrast between computer-generated speech and human speech. Computer-generated speech is monotonous, conveying less meaning than human speech, and our attention wanders when we hear it. Most animal vocalizations depend on varying contributions of emotions and cognition. The food calls given by many species (such as chimpanzees, macaque monkeys, and chickens) are not monotonous and it is not the case that identical vocalizations are produced every time food is found. Instead, they vary according to how much food there is and the quality of that food. Chickens, for example, produce food calls at higher rates when the food is of the preferred kind, but other aspects of the call are varied in other species. The information about quantity and quality may be generated by the emotional state of the chicken that is producing the calls, since both more food and food of better quality may increase the bird's excitement.

At the same time as they express the emotions, signals can be referential (which requires one form of cognition), and they can be emitted or suppressed depending on the presence or absence of an audience, or on other external factors. The relative importance of emotional versus referential processes varies with the particular call. For example, the leopard alarm call of vervet monkeys appears to have more emotional content than either the eagle alarm call or the snake alarm call—the leopard alarm call has been observed to occur occasionally in aggressive social interactions and sometimes when a raptor swoops down at the monkey, whereas the eagle and snake alarm calls have never been heard unless the specific predator to which they refer is present. Each of the latter two calls, therefore, has a unitary meaning, whereas the leopard alarm call appears to have more than one meaning.

Joseph Macedonia and Christopher Evans (1993) have, however, reasoned that the leopard alarm call does not simply signal the monkey's level of excitement or fear, as in the case of the whistle calls of ground squirrels, because the same leopard call is produced whenever the monkeys see a leopard regardless of what degree of threat it actually poses—whether the leopard is asleep, hunting, attacking, moving away, or approaching. It could, of course, be argued that a leopard causes maximum levels of fear irrespective of what it is doing, and that may be why there is no variation in calling.

So far, most research on the referential use of vocalizations in animals has focused on signaling about the presence of predators or food, but much of the communication in animals must be concerned with social relationships. Although survival depends on alerting conspecifics to predators and food, social interactions are an equally important aspect of an animal's life. It follows that a considerable amount of communication must occur about social situations, but virtually nothing is known about these forms of communication. Certainly, Cheney and Seyfarth (1985, 1990) have shown that vervet monkeys are aware of the social relationship between a mother and her offspring: when an infant vocalizes in distress, other monkeys turn to look at the mother of the infant rather than going to its assistance themselves. This is an example of active suppression of a response to a signal by monkeys who are not related to the infant. It shows that social signaling depends on the social context.

There are many other ways in which vervet monkeys, and other species, may communicate about social situations in an active manner, but unfortunately we know nothing about this potentially rich field of communication. As Sue Savage-Rumbaugh has said, apes may be less interested in communicating about objects than are humans and more interested in communicating about social matters (Savage-Rumbaugh and Lewin, 1994).

We are far from understanding communication at the social level, but it is reasonable to say that, although emotional states may be an aspect of social communication in animals, communication may also be generated by cognitive processes.

ANIMALS THAT UNDERSTAND HUMAN LANGUAGE

It would help us to find out for certain whether animals communicate intentionally if we could ask them what they intended to communicate and they could reply using communication signals that we could understand. There are two potential ways of achieving this two-way communication:

either we could learn to use the communication signals of the animal species we wished to study or we could teach the animal to use some form of human language. Since we have not yet been successful in understanding more than rudimentary aspects of animal communication signals, the latter has presented itself as the best option. Apes have been taught to communicate with humans using American sign language or by pointing to symbols that represent words. They have not been taught verbal communication using spoken English, for example, because the vocal apparatus of apes is very different from that of humans and does not allow them to make the same range of vocalizations that we do. This does not mean that apes' vocal abilities are limited—they can and do use a range of complex calls, with a vocal range extending to very high frequencies. Birds can produce the same range of sounds as humans, and they can be taught to communicate with humans using vocal signals, as was a parrot called Alex.

We will discuss intentional communication in apes first. Allen and Beatrix Gardner trained several chimpanzees to communicate with humans using American sign language, beginning in 1966 with one called Washoe (see Gardner, Garner, and van Canfort, 1989). The chimpanzees learned to use signs to refer to objects and individuals, and all of them acquired vocabularies that allowed them to express requests, such as "Icecream, hurry gimme" (to use the Gardners' translation), "You tickle me Washoe," "Please flower," "Please blanket out" (requesting a change in location of a blanket then in the cupboard), "You me out" (a request for the human observer and chimpanzee to go outside), and "Open help" (requesting assistance in opening a lock or a bottle). By the chimpanzees' frustrated behavior when these requests were not honored, compared with when they were, it was clear that these were intentional forms of communication. The chimpanzees also announced when the next activity in the daily routine should occur with statements such as "Time vacuum," "Time toothbrush," and "Time Dar out" (Dar being the name of one of the chimpanzees). This announcement of a pending event is an aspect of awareness of the future that indicates intentionality. Emotion entered into the signing—more emotive events evoked more signing—but cognition was obviously a major aspect of their communication.

To convince critics that the chimpanzees were expressing genuine requests and were coming up with answers to questions by use of their own powers of cognition, it was necessary for the Gardners to prove that the chimpanzees were not using subtle cues given inadvertently by the humans caring for them. By responding to cues produced by the humans in their presence, the apes could appear to be communicating intelligently and intentionally but would merely be performing some sort of clever mimicry. In other words, they might be similar to Clever Hans, the horse that was once thought to be able to read numbers written on a board and to count them out by tapping his foot on the ground. Later it was found that the horse used subtle cues that his owner supplied unknowingly, such as the blink of an eyelid when the horse tapped the required number of times. Clever Hans could not perform the task when his owner was not present.

To test whether a similar use of cues might be occurring with the chimpanzees, the Gardners designed an experiment in which the chimpanzees had to name objects shown to them on a video monitor. Their responses were recorded by a human who could not see the screen and did not know what the chimpanzees were observing. There was no human who knew what was on the television screen present in the room with the chimpanzee. In this controlled experiment, the chimpanzees were able to name objects accurately. Therefore, their use of sign language was self-generated and not some form of mimicry or associative learning.

The chimpanzees also used the sign language they had learned to tell humans things they did not already know. For example, when very young, Washoe dropped one of her toys into a hole in the inside wall of the caravan in which she lived. That night, when Allen Gardner visited her, she attracted his attention to a part of the wall below the hole and signed "Open, open" many times over. From this communication Allen deduced what had happened and retrieved the toy. Washoe had used sign language to communicate something really new to a human. This shows genuine communication with intention. Again, there is no question of the chimpanzee's having communicated merely by reading subtle cues given by a human. Although this was claimed rather vehemently by several researchers in the field at one time, a complete analysis of the data accumulated by the Gardners shows that this narrow interpretation is most unlikely to be correct. Moreover, more recent findings by other researchers who have trained apes to use language support the conclusion that apes can learn to communicate with humans intentionally, creatively, and intelligently.

Sue Savage-Rumbaugh has trained chimpanzees and a bonobo (a rare species of chimpanzee, *Pan paniscus,* also known as a pygmy chimpanzee) to communicate with humans by pointing to symbols on a board (lexigram symbols) (see Savage-Rumbaugh and Lewin, 1994). She has said that while the sign-language–trained chimpanzees used their acquired language mainly to manipulate humans, the symbol-trained chimpanzees seemed to have more of a two-way communication with humans. This claim needs to be proven, but if it is correct, manipulation by the signing chimpanzees might stem from the fact that they were trained by being given small food rewards when they signed correctly, and thus they would associate signing with getting a food reward from humans. The other factor that might be important is that humans communicated with the symbol-trained apes using spoken language, not sign language, which is generally used to communicate with chimpanzees trained to use sign language. The combined use of speech by humans and symbols by the apes might have facilitated the human–animal exchange because humans could speak to the apes directly using their natural form of communication. Whatever the reason, the symbol-trained apes have impressive two-way communication with humans, and they are able to use that communication to refer to events that have occurred in the past or to talk about other individuals not present at the time the conversation takes place. This is clear referential use of communication.

Not only is meaning—referential use of symbols—important to these apes, but so is syntax, the grammatical word order in sentences. This was discovered by Savage-Rumbaugh in her work with the bonobo Kanzi (Savage-Rumbaugh and Lewin, 1994). Kanzi learned to communicate, by pointing to symbols, by being present at an early age when his mother was being taught to use them. He learned to use the symbols to generate language in much the same way that a human child acquires language. He also acquired the ability to understand spoken English. In a sense he is now trilingual, because he is able to understand spoken English, to use symbolic language, and, most likely, to use his own "chimpanzee" mode of communication. This is more than we expect of the average human child.

Kanzi also learned to understand the syntax of English, as Savage-Rumbaugh was able to show in the following experiment. Kanzi was given instructions via headphones by a person in another room who could not see him. In the same room as Kanzi was another person who did not know what instruction Kanzi had received and who recorded his behavior. Kanzi was instructed to perform a task in pidgin English ("Go get orange testing room") or in syntactically correct English ("Go and get the orange from the testing room"), and the rapidity of his responses was recorded. The results demonstrated that he responded more rapidly and more effectively when the syntactically correct instruction was given than when pidgin English was used. Therefore he has acquired understanding of not only the meaning of words (semantics) but also the structure (syntax) of English language. The symbolic language by which he has learned to communicate with humans does not permit this expression of syntax, but he does process and respond to syntax. In fact, Kanzi can understand numerous sentences in spoken English.

This remarkable demonstration of Kanzi's ability to understand human language shows that apes possess the ability to process language and that they may use this ability in their own vocalizations or other forms of communication. It even raises the possibility that other species that live in close contact with humans understand what humans are saying even though they cannot themselves speak. In his work with dolphins at the University of Hawaii, Louis Herman and his colleagues recognized this possibility of comprehension in the absence of audible or visible production of signals, although in this case he was considering the dolphins' ability to understand the gestural "language" of humans rather than speech (Herman, Pack, and Palmer, 1993). Dolphins can follow complex commands presented to them as gestures asking them to perform various acts in sequence, even though they have not been trained to produce vocal or other communication that can be understood by humans. In saying this, we must not overlook the dolphins' own complex vocal and other forms of communication, which might also share aspects of human language.

It is possible that many species that live in close contact with humans acquire some comprehension of both the semantics and syntax of human language even though they cannot produce it. We all know that dogs, for example, understand simple commands, but we might also speculate that they understand much more of the conversations we have in their presence, and the same may be true of pet birds. In fact, a study by Millicent Ficken, Elizabeth Hailman, and Jack Hailman has shown that chickadees (*Parus sclateri,* an avian species in Mexico) sequence their different calls in particular ways according to rules and the context in which the calls are given. They thus use a simple form of syntax, as the researchers state (Ficken, Hailman, and Hailman, 1994). It is probable that many other examples of syntax will be found in the communication systems of animals, and it need not be found only in vocal communication.

Of all the species that could have been chosen for the research on teaching human language to animals, apes were not a surprising choice. Apes are closest to humans genetically and in terms of evolution; so it was considered they would be more likely to be able to learn to communicate using language than any other species. Irene Pepperberg, however, saw potential in training a species far removed from humans—the Grey parrot *(Psittacus erithacus).* Parrots have an advantage over apes because they can mimic human speech vocally and might be able to communicate directly without the need for an interface of signs or symbols.

In her laboratory Pepperberg (1990a, 1990b) began by training a parrot called Alex. The training had to differ from the usual way in which parrots are taught to mimic speech. Instead of mindlessly repeating words or phrases over and over to the bird quite out of context and therefore without particular meaning to the bird, she and her students engaged in simple but meaningful interactions in front of Alex. For example, one person would ask, "Where is the key?" and another would hold the key up with a reply such as "Here is the key." The first person would then ask, "What color is the key?" and the other person would state the color, and so on with objects of different colors, shapes, and textures. When Alex began to use words, he was given the objects that he asked for, and the humans also rewarded him by telling him he was a good bird.

With this training, Alex has learned to name up to 100 objects and to answer questions correctly about their shape, color, and texture. He can also count, and when presented with an array of objects of various shapes and colors on a tray, he can say how many of the objects are, for example, green triangles or blue four-corners (by which he means cubes). Alex also expresses desires, such as "I want peanut" or "Come here." In all aspects of his communication, he performs as well as the language-trained apes, a fact that supports our suggestion that many species may be capable of understanding aspects of human language. This comment aside, the relevant point about the research with Alex is that he uses his acquired vocabulary to communicate intelligently with humans. He is not simply emitting signals mindlessly, out of context or unintentionally.

CONCLUSION

In 1975, the primatologist David Premack asserted that whereas humans have both affective and symbolic communication, all other species, except those tutored by humans, have only affective communication (Premack, 1975). By affective communication he meant communication about emotions in an uncontrolled way. At the time he wrote, apes had already been taught to communicate using sign language (Premack himself had been part of the research program), and their abilities to communicate symbolically were known. Instead of extrapolating this knowledge to communication by species using their own species-specific patterns of communication, Premack saw the language-trained apes as exceptions that had acquired something extra as a result of their contact with humans. The more recent research of Marler, Evans, and colleagues on vocalizations in chickens discounts Premack's claim. They have shown that alarm and food calls are not simply produced automatically without control (Marler and Evans, 1996). We might therefore conclude that the apes who have been taught to communicate using signed or symbolic forms that humans can understand tell us something important about their species and, in that respect at least, are not exceptions.

The research on Kanzi and his ability to comprehend the syntax of spoken English has led us to suggest that many other species may have similar abilities despite the fact that they cannot speak to us or communicate by signing or using symbols. We would go a step further and suggest that the ability to understand the syntax of spoken English indicates that bonobos, at least, must communicate using their own species-specific signals (vocal and gestural) in ways similar to the ways humans use language. We base this statement on the fact that a species that can comprehend human language is also likely to have similar processing capabilities that are used for its own forms of communication; in turn, this means that the species must produce language-like communication. The fact that language-like production of communication has not yet been found in animals tells us only that far too little research on natural communication has so far taken place—it does not tell us that it does not exist.

In fact, detailed examination of the vocalizations of different species frequently reveals that humans are unable to distinguish between calls that actually differ from each other. In other words, we may not hear differences that the animals hear. Many years ago, this was found to be the case for the most common call of Japanese macaque monkeys *(Macaca fuscata),* known as the "coo" call. The monkeys make "coos" in a variety of social situations, and although these all sound the same to human listeners, detailed analysis revealed that the calls differed in each situation. More recently, the same has been found for the trill calls of spider monkeys *(Ateles geoffroyi):* although all trills sound the same to us, spider monkeys can tell exactly which individual made the call. These examples should indicate to us that there is much more in the vocal communication of animals than we hear or understand.

Chapter Four

COMMUNICATION IN BIRDS

Birds have inspired human imagination. To fly like a bird is a dream as old as Greek mythology and the desire of Icarus to fly away from Crete on wings made of wax. The white dove has come to symbolize peace; birds also symbolize freedom. Bird feathers have been used to signal special powers or to confer a special status on the human wearer. Birds feature in many human dances—many cultures have prided themselves on being able to mimic birdsong and bird displays. Human fascination with birds may also arise from having something in common with birds—humans and birds share a strong investment in communication by vocalization. In fact, the complexity of song and communication systems developed by birds and by humans has no equal among vertebrates, except for whales and dolphins.

Birdsong had been studied and described long before scientists took a scholarly interest in it. Today, the study of birdsong is a substantial field in its own right. It is studied for the sake of learning about its communicative value, but also because it is aesthetically pleasing. It may be described in terms of its structure as well as its function. Researchers may be interested in the acquisition of song or in how and where song is produced. Ethologists are interested in birdsong in relation to questions of territory or reproductive strategies.

In evolutionary terms different species of birds may be as far apart from each other as ungulates are from humans. The first bird evolved in the Jurassic period, although most ancient bird species evolved later, in the Cretaceous period. Millions of years separate the appearance of the various species (Feduccia, 1996). For instance, the first known occurrence of some flightless birds, including species of game birds and waterfowl, may have been close to 100 million years ago, separated from the appearance of parrots by about 10 million years. Owls evolved about 60 million years ago, about 30 to 50 million years earlier than songbirds.

Songbirds and most other birds of prey were among the "newcomers," appearing in the Tertiary period as recently as a mere 5 to 30 million years ago. Albatrosses, frigatebirds, penguins, and petrels evolved earlier. Thus, when humans began to evolve about 4 million years ago, the air, the ground, and the waters were already occupied by winged and beaked species.

Some people have thought that all animals that have wings and lay eggs are the same kind of creature, but birds' evolutionary distance from one another and their differences in behavior make this as absurd as saying that mice and tigers are similar. But it is not just for reasons of appearance that birds have been seen as a unitary set of species; the history of ideas has also played a role. Descartes's notion that only humans are "complete" beings by virtue of their ability to think had particularly bad repercussions for birds. A false impression was created that birds are essentially like mechanistic toys. Likenesses of birds have been used as colorful decorations in living rooms or as self-propelled music boxes on mantlepieces, just to adorn human dwellings, with little thought of the live birds.

Some birds have evoked a negative image of evil or death. Think of the crow on the witch's back. Vultures are a symbol of death, and any haunted house worth its reputation has birds flying from it, dark and menacing with their sharp beaks and claws. Such associations gave rise to the links between birds and bats and prehistoric monsters. This negative imagery of birds contrasts with the positive association with their flight and songs.

But even positive images of birds have not led to the abandonment of the view that birds are less capable of higher cognition than mammals. The prevalence of this view has a number of consequences for studies of communication in birds. It influences what we prejudge as being the capability of birds and affects what we discern as human observers. In certain avian species, basic vocal signals may be innate and automatic. But in more complex avian species, such as the psittacine group (parrots, cockatoos, budgerigars), corvids (crows, ravens, jays), and the Cracticidae (Australian magpies, currawongs, butcherbirds), studies have shown that many of the vocalization skills are learned behaviors. Mastery of these skills is partly responsible for success in finding a partner, breeding, and acquiring and holding on to territory. There is nothing automatic about the production of vocalizations even in chickens *(Gallus gallus)*, even though their vocalizations are simple compared with those of songbirds. Hence in studies of communication in birds we may often be dealing with learned, complex vocalizations and very complex social interactions.

We are only just beginning to understand the complexity of bird communication. Researchers who have shown that a variety of birds are capable of complex communication have fought traditional views. Irene Pepperberg's (1990a, 1990b) research on communication by her Grey parrot Alex is one example. Alex communicates with the researcher using English words. He can count and discriminate shapes, colors, and objects. He can understand commands and express wishes. Pepperberg's research suggests that Alex has learned to use English words to communicate in a comprehensive way, not simply by mindless association of certain words with certain events. These capabilities appear to be the result of thinking (or consciousness) rather than automatic responses. Alex the parrot may well be on a par with the great apes in his abilities to communicate and reason. Many other avian species may have abilities similar to those of Alex. One famous Australian corella, a particularly argumentative species of cockatoo, has learned to argue with and even shout at its owner in human language (BBC, 1996). The studies by Konrad Lorenz (1966) on corvids (European ravens) and Kaplan's recent studies on the Australian magpie (1996, 1999) show similar complexity of communication.

METHODS OF COMMUNICATING

When we speak of communication in birds, it is customary to look at the relative importance of visual, auditory, and olfactory communication for individuals of a given species. In some classic studies it was found that, in relation to other means of communication, acoustic signals are of prime importance. Many years ago this was shown to be the case in the domestic hen's recognition of her chick. When a transparent bell was placed over a small chicken, preventing the hen from hearing its calls, the hen paid no attention to the distressed chick. A turkey deprived of auditory cues also failed to recognize her own offspring, and consequently attacked and even killed them, as she would any intruder. Hence, in some contexts and in some avian species, acoustic signaling is dominant over visual signaling, but most communication in birds makes use of more than one sense simultaneously, especially the visual and auditory senses. Many species of birds perform visual displays while vocalizing.

Recognition of individuals need not involve sophisticated vocalization patterns but may require excellent hearing. The acute hearing of many birds may also be used to find food. For instance, the Australian magpie *(Gymnorhina tibicen)* locates its food largely by sound. Its hearing is so good that it can locate scarab larvae moving in the soil several inches under the surface. The Australian tawny frogmouth *(Podargus strigoides)* could theoretically find its food if blindfolded. It can hear the rustle of beetles and cockroaches in the undergrowth from the height of a tree branch. We might expect acoustic signals to bear characteristics relevant to the bird's social and ecological environment, and also to its hearing capacity, but this is not necessarily so. Australian magpies have very intense, high-amplitude calls and songs, which seems extravagant given their exceptional hearing. By contrast, the tawny frogmouth uses low-amplitude and low-frequency sounds to communicate. Tawny frogmouths, unlike magpies, are nocturnal. In the stillness of an Australian bush night, the repetitious hoot of the tawny frogmouth can be heard for miles.

Visual Signals

Although birds use vocalizations extensively for communicative purposes, they are by no means their only way of communicating. Visual communication is used widely by birds, requiring suitable eyesight to perceive the visual signals. A study by Patrice Adret has shown that visual stimuli (in the form of video images) have reinforcing properties in zebra finches *(Taeniopygia guttata)*, although the study allowed auditory cues as well (Adret, 1997). Merely showing the head of another male zebra finch on screen roused the experimental bird to song. Bengalese finches *(Lonchura striata domestica)*, investigated by Shigeru Watanabe, were found to rely predominantly on visual cues for discriminative behavior. The auditory signals in his experiments provided purposely ambiguous information and in those cases the bird's attention switched to visual signals that were not ambiguous (Watanabe, 1993). Australian magpies and tawny frogmouths are now also known to use visual signals to communicate. This was established in an experiment by Kaplan. For thirty days Kaplan wore the same clothes while she fed the birds. On the thirty-first day, she exchanged the feeding clothes for others with different colors and designs. Even though the auditory cues remained the same, birds of both species showed fear responses. On day 32, Kaplan carried the regular feeding clothes into the aviary and changed from the new to the regular feeding clothes in front of the birds. The fear responses disappeared in both birds on completion of the change. Interestingly, however, the magpies adapted to allow any form of clothing from then on, but the tawny frogmouths continued to show fear responses even with slight variations in the clothing. Of course, a change of clothing may be perceptually difficult to accommodate, because birds rarely change plumage color and patterns other than when they mature from nestling to adult or, in some species, when they change plumage with the seasons, as do the partridge in Europe and the male superb fairy wren in Australia.

It is difficult to speak about visual perception in birds in general terms. There is a tremendous diversity in optical designs and retinal structures across avian species. Some species even have infrared or ultraviolet vision. Owls and a variety of other nocturnal species (such as owlet nightjars) can see at very low intensities of light. Diurnal birds of prey have probably the best long-distance sight and visual acuity of any species. Most bird species have limited movement of the eyeball but this is compensated for by great flexibility in head movements. The eye of the barn owl, for instance, is fixed rather firmly in its socket, but the head can move 270 degrees, both vertically and horizontally. There are only two bird species so far investigated that show extensive movement of the eyeball. One is the bittern, which in its "freezing" position (head and beak up, neck stretched) can turn the eyes forward and downward to see below its beak and straight ahead in binocular vision. Another is the snipe, which can turn its eyes upward to watch a bird overhead without moving its head at all.

The eyes of birds are often at the side of the head, and therefore a good deal of visual information is obtained in monocular vision. This provides a large visual field, including areas above and, in some species, behind the head of the bird. This kind of vision is of great advantage for survival, but it is not clear whether it serves any communicative function. Certainly, birds perform lateral (or broadside) displays that may require use of the lateral field of vision.

Like mammals, avian species have a wide range of body postures available for signaling a message by visual means. Head bobbing, arching of the neck, extending of the wings outward, and certain sorts of running, stomping, and crouching postures may be used in both agonistic (threat) and courtship behaviors: the same posture can have embedded in it the potential for both flight and attraction. Many courtship rituals rely on rapid changes in body posture.

One of the best known and most dramatic courtship rituals that relies largely on motion and body posture is performed by the grebes (*Podiceps* spp.) as a dance on water. The sequence is rather complex: In horned grebes *(Podiceps auritus)* the male "bounces" forward and dives several times: then both male and female rise to full height by treading water, facing each other in what is sometimes referred to as a "penguin" display; they continue to dance in that posture until finally they swim apart. The village weaver male *(Ploceus cucullatus)* uses a wing- and head-pointing display to attract a female's attention not only to himself but to the nest he has built. There are many bird species that use dance or ritualized movement as a component of their courtship display. Lyrebirds (*Menuridae* spp.) are famous for their dancing as well as their vocal displays (see Robinson and Curtis, 1996).

Plumage Signals

Feathers are often used for signaling. Although plumage color is not a universal factor in recognition of the sex of a bird, it plays this role in a large number of species. Males of many species use it to attract a mate. Recognition of sex, in some species, may occur exclusively through visual cues—plumage color or eye color, as Glenn-Peter Saetre and Tore Slagsvold from Oslo found in experiments with caged pied flycatchers *(Ficedula hypoleuca)*. When they painted a pied flycatcher female in the colors of the male, all other males treated the bird as if it were male. A male painted as a female was treated by the others as a female. This identification was maintained even when the song of the male was played in conjunction with presentation of the male bird painted as a female (Saetre and Slagsvold, 1992). It is worth noting here that some male pied flycatchers naturally have plumage coloration that is closer to that of the female. In free-ranging birds, males treat such birds as if they were female and may even engage in courtship rituals for their benefit. Males equipped with a plumage color that mimics that of a female can accrue territorial advantages. They may invade a territory without encountering the aggression of a competing male and may succeed in staying.

Apart from sexual recognition, plumage color and patterns may signal such things as individual identification, dominance status, and mating readiness. Plumage may also provide a sign stimulus (Konrad Lorenz's term) that can lead to attack. For instance, the red breast of the male European robin functions as such a sign stimulus that leads to attack by other males. Even stuffed models placed on a branch provoked attack when the breast was red, but not when the red was missing.

While sexual recognition and individual identification as a result of plumage are passive, pregiven signals that are genetically determined, dominance status and mating readiness require some additional, active social communication to get their meanings across. Birds of paradise, for instance, go to extraordinary lengths to display their plumage. As mentioned earlier, the male Victoria's riflebird *(Ptiloris victoriae)* will choose a sunny, exposed part of the rainforest and rhythmically display tail or wing feathers, performing a fascinating dance, with colors flashing in the sun, that will attract a female to come close for inspection. The male will then proceed with his display, but this time he half folds his wings around the female (without touching her) in quick succession, first the left and then the right, in such a way that the female becomes almost a captive in the courtship ritual.

Perhaps the most spectacular use of feathers in signaling is shown by the peacock with tail feathers fanned out like a wheel, shimmering with each new turn of the body. As well as its iridescent green and blue colors, the peacock's tail has hundreds of eyespots, patterns that mimic eyes, all appearing to be looking inward toward the body of the peacock. In Chapter 1, we saw that eye-like patterns (ocelli) are used by some moths and butterflies to direct the attention of predatory birds away from the body and to the less vulnerable wings, or even to startle the bird long enough to escape. The peacock uses the eye-like patterns for intraspecies communication to attract a female during courtship.

An eye-like pattern is used for the same purpose by males in one of the 43 species of the birds of paradise, the bluebird. The pattern is hidden at the abdomen and surrounded by magnificent long bright blue feathers. Only during courtship do these blue feathers with their eye-like markings (black and shiny bright red) come into full display. For this to happen, the bird needs to hang upside down on a branch and fan out all the blue feathers to expose the eye-like pattern. These then hang over the chest. He swings them rapidly to and fro while emitting rasping, rhythmic, and mesmerizing percussion sounds in quick succession.

A most unusual and complex form of visual display occurs in bowerbirds (*Chlamydera* spp.). Here the display of feathers has been replaced by decorations external to the bird. We could almost speak of tool use. Instead of, or in addition to, parading bright or striking plumage to a prospective female, males build a bower. The bower may be decorated with all manner of objects of similar colors, depending on the species' preferences. During courtship, the male displays plumage and may vocalize and even dance, but there is always the additional element of a stage, uniquely constructed specifically for the purpose of attracting a female. Like the lyrebirds, male bowerbirds clear an area on the forest floor for dancing. In addition, the males of most bowerbird species build a structure that is of no use for raising young but is an integral part of their courtship display. Mating success is linked to the bower and the entire display, including vocalizations, dancing, and construction of the site. For the best performers, the enormous effort pays off by giving them access to many females.

Signals issued by feather posture alone have rarely been studied systematically, yet they may be quite important ways of communicating visually. Many birds fluff their feathers in a certain way when they are ill but they may also raise their feathers as a warning signal. Tawny frogmouths can raise all their body feathers simultaneously to make themselves look menacingly larger than they are. This display is not necessarily accompanied by a vocalization, but it always precedes an attack and appears to be used in territorial disputes among conspecifics. In interspecies interactions, tawny frogmouths seem to shrink their body size by sleeking their feathers down as close as possible to the body and by stretching their necks. The bird then gives the appearance of a branch, a camouflage that works well against a gum tree (see Figure 4.1).

Facial Expressions

It is equally possible to attribute communicative importance to the facial expressions of birds. The idea that birds have "facial expressions" is quite foreign to many people and there has been no systematic work done on this aspect of avian communication. The concept that a bird has a "face" may seem strange, but that is largely because humans have linguistically claimed the "face" as something uniquely human, a feature that bestows individuality. (There are now a few select mammalian species to whom we grant individuality and thus a face). Although it is recognized that many avian species express individuality in their vocalizations, it is usually not accepted that birds do so in their appearance. But individual birds do "look" different, and they look different in different

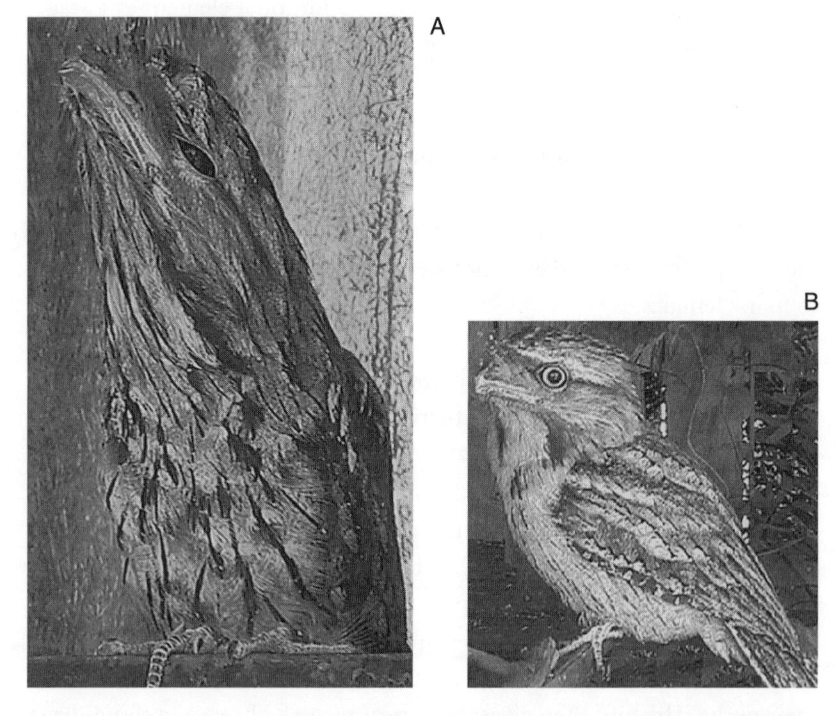

FIGURE 4.1 Postures of the tawny frogmouth. A: Feathers on head and back sleeked down and neck and body extended in a camouflage posture. B: Neutral posture adopted by the same bird. (Photographs by G. Kaplan.)

contexts. Facial expression is achieved either by movements of the beak or by independent positioning of feathers on the chin or above the beak, on the ear coverts, on top of the head (the crown), at the nape of the neck, and in some species also by the bird's moving the feathers above the eyes independently of the other feathers.

Like primates, birds have open-mouth displays, really open-beak displays, which, together with other body signals, can be used in fear or threat displays. Many species use a variety of open-beak displays; in the tawny frogmouth, the open beak displays the inside lining of the large oral cavity, which is a striking light-green color. This effectively emphasizes the enormous size of the beak and makes it look more ominous than it actually is. Several species of birds open their beaks as a threat, usually without vocalizing but sometimes augmenting this display with hissing or breathing sounds. Galahs and many other psittacine (parrot) species use open-beak displays accompanied by shrieks, hisses, or exhaling-air sounds. The barn owl *(Tyto alba)*, for instance, a bird that is rarely heard, emits an exhaling-air sound in warning while the beak is half open and then sharply claps the upper and lower parts of the beak together several times, often without the slightest change in body posture or feather composition.

In galahs and other crested cockatoos, movement of head feathers is very easy to detect, even from some distance. The crest goes up not just in alarm but in states of friendly arousal. The feathers that flank the beak (the ear coverts) can be ruffled to express anger and possible attack. Lowering or flattening of feathers is usually associated with fear, but such a display commonly involves the whole body rather than just the head. Birds often show "cuddly" and babyish behavior by fluffing the feathers above and below the beak, a phenomenon readily observable in Australian magpies. For close, conspecific interactions, these facial expressions are powerful signals emitted with a minimum expenditure of energy.

Smell and Touch

Compared with auditory and visual communication, relatively little is known about communication by smell and touch in birds. We know that many bird species preen each other, usually as an exercise in bonding and reassurance. For some avian species, particularly parrots, preening and tactile responses are very important in social interactions. In some species, such as the red wattlebird *(Anthochaera carunculata)* and the Australian magpie, newly hatched nestlings that have not yet opened their eyes will not defecate until they feel the vibration at the nest indicating the presence of a parent. They then lift their cloacal region toward the edge of the nest and the parent takes the firm feces into its beak and carries the waste out of the nest. Although this is not exactly a form of tactile communication, the tactile signal of the parent elicits the response. Later in the development of magpies, the parent may actually prompt defecation by tapping its beak directly on the offspring's cloacal region (observed by Kaplan).

Olfaction in birds is less well developed than the other senses but it is not absent. Olfactory cues have been shown to play a role in food selection in a number of species (as has been summarized by Malakoff, 1999), but the studies in this field are limited and it is not known whether odors are used to communicate between individuals. The sense of olfaction is unusually well developed in the New Zealand kiwi *(Apteryx australis)*, which locates its food by sensing odors (Wenzel, 1972), so it is probable that olfaction is also used for communication in this species. The same may be true of other species of birds that are known to locate food by its odor. These include the turkey vulture, *Cathartes aura* (Stager, 1967), a number of shearwaters and petrels (Grubb, 1972), the common raven, *Corvus corax* (Harriman and Berger, 1986) and the starling, *Sturnus vulgaris* (Clark and Mason, 1987).

The tawny frogmouth defecates as a deterrent when feeling threatened, as in cases of mobbing attempts by other avian species. The bird will fly close to the animal to be deterred and deliberately spray large quantities of its extremely pungent excrement over or near it. Tawnies are the skunks of the air, and their warning scent, if dropped on fur or skin, is difficult to eliminate. Usually, this warning signal is reserved purely for other species, and it is not clear whether tawny frogmouths themselves can actually smell their own droppings or perceive the intensity of the odor of their excrement.

Even if auditory and visual cues are likely to be the most important signals in avian communication, it can at least be said that no single sense functions entirely in isolation. Courtship displays in birds are a good example. Auditory messages are usually accompanied by visual displays that can be very elaborate, involving motion and even "dance." Some tactile contact may also be part of the ritual (neck touching, beak fencing, or, more indirectly, the exchange of gifts).

HOW VOCALIZATIONS ARE PRODUCED

Vocalization depends on appropriate centers in the central nervous system. In birds, there are a number of specific, so-called sound-emission sites, some of which are conspicuously different from those of other vertebrates (mammals, including humans, reptiles, and fish). The chief sound-producing vocal organ of a bird is the syrinx. Although avian species also have a larynx, like humans, Rod Suthers (1990) and others have thought that the larynx plays no significant role in sound production. However, researchers are still debating whether supersyringeal structures, such as the trachea, larynx, tongue, and even the upper and lower mandibles, play a role in modifying sounds. The tongue may be important in psittacine species. Recent work by Dianne Patterson and Irene Pepperberg (1996) on American English vowel production in the Grey parrot has shown that this parrot can produce vowels of striking similarity to human vowels despite the very different anatomy of the psittacine vocal apparatus (lack of teeth and lips, for instance). Indeed, many parrots can produce such vowels, as a sonogram of galah "speech" shows (Figure 4.2).

There are several important differences between the avian and the human vocal apparatus. The most obvious one is the location of the main sound-producing organ. The human larynx is situated in the neck, and hence is close to the mouth. The avian syrinx, by contrast, is located well within the body of the bird. It sits at that part of the trachea (windpipe) where the bronchial branches split and go to the lung on one or the other side of the body. Thus a bird has two airstreams impinging on its vocal organ rather than one, as in humans. The onset and termination of vocalization (called phonation) is usually controlled by the syringeal muscles that open or close the lumen (airway) on each side of the syrinx.

The syrinx is an organ that varies in complexity from species to species. Although the precise mechanisms of sound production are not fully known, it is thought that voiced or whistled song originates from vibration of the medial tympaniform membrane. The syrinx has internal medial tympaniform membranes that are housed within the interclavicular sac, an air sac in the pleural cavity. In that location the membranes are sensitive to the air passing through from the lungs, and they are controlled by the syringeal muscles and by air pressure surrounding the membranes. The elasticity and complexity of the membranes may determine the quality of sounds.

Songbirds have a very complex syringeal system, in which the syringeal muscles and the internal membranes interact to produce nearly pure tones (single-frequency tones, similar to human whistles) and also, as in the lyrebird, parallel notes, seemingly played on two instruments at once. The latter sounds are produced from both sides of the syrinx at once, as we will discuss below.

In the Australian kookaburra, with its loud and raucous call, the syringeal muscles are barely developed. By contrast, the syringeal muscles of the Australian magpie are very noticeable. It is possible to trace the development of song in an Australian magpie in relation to the development of the syringeal muscles. Full song is produced only when the syringeal muscles are fully grown.

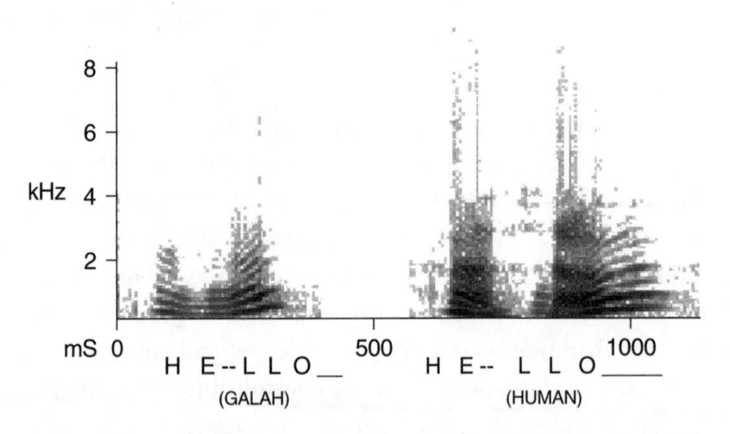

FIGURE 4.2 Mimicry of human speech by a galah *(Cacatua roseicapilla).* Time on the Y axis is in milliseconds. Although the "hello" produced by the galah is of shorter duration than that spoken by the human, the patterns of the two vocalizations are almost identical. The fundamental notes and the overtones are very similar. (Sound spectrograms produced by G. Kaplan.)

It is possible for a songbird to produce sounds from different sources at the same time—to use both sides of the syrinx simultaneously or independently. Suthers confirmed this in 1990 for the brown thrashers *(Toxostoma rufum)* and gray catbirds *(Dumetella carolinensis).* He found that in both species the frequency range of sound contributed by the right syrinx was higher than that of the left syrinx, and that phonation was frequently switched from one side to the other, not just in between syllables but within a single syllable. Simultaneous use of both sides, at least in the species they examined, resulted in syllables that are "two voiced," syllables that are not harmonically related and are of different amplitude modulation. This means that one side of the syrinx is not dominant, as is the case with canaries. In addition, some birds utilize air reservoirs in the chest as resonance chambers for the production of sound. These may be used in conjunction with the syrinx, or even in the absence of a syrinx, as for instance in bustards, emus, and cranes, which have clavicular and cervical air sacs. The sounds these air sacs produce are hollow and of low frequency, like the sound of a drum being struck under water.

The sound repertoire is not exhausted at this point. Some species use beak clapping to communicate. Beak clapping in storks, some owls, in the three frogmouth species, in noisy miners (*Manorina melanocephala*), and in Australian magpies is used as a strong and aggressive warning signal to other species. Some species also generate auditory signals by pecking an object, such as a tree, to indicate territoriality, and woodpeckers use such signals for sexual communication. The male musk duck in eastern Australia *(Biziura lobata)* produces as part of his courtship ritual an odd "plonking" sound of his feet in the water.

Wing flapping and wing beating, as in wood pigeons and crested pigeons, may function as warning signals. The sound of wing beating by a crested pigeon *(Geophaps lophotes)* is a high trill that can be heard some distance away, and since not every flight motion produces this shrill sound, we suspect that wing beating in crested pigeons is used for communicative purposes. William Thorpe and Donald Griffin (1962) found that the flight sounds of some small birds contain ultrasound. They are therefore not audible to the human ear and probably also not to birds such as owls that prey on these small birds, but they are certainly audible to bats and other vertebrates with ultrasonic abilities. A large number of songbird species, nearly all tits (*Parus* spp.) for example, show a behavior called "wing quiver," which is said to have communicative function. The wing quiver is caused by a vibratory movement of the wings, mainly the wingtips, at a sound frequency of about 15 Hz. In the black-capped chickadees *(Parus atricapillus),* wing quivering usually occurs in front of the nest hole before the bird enters the nest, and Marcel Lambrechts and his colleagues concluded from their observations that wing quivering functions as a request or invitation to the mate (Lambrechts, Clemmons, and Hailman, 1993).

A most unusual way to produce sound, for a bird, is tool use. The male palm cockatoo *(Probosciger aterrimus),* found only at the very tip of Australia's tropical north and in New Guinea, fashions a stick to a manageable length. He then holds it in one foot and drums the stick on a tree while emitting very high-pitched but not very loud shrieks, dancing at the same time and swaying his head. With this triple activity—swaying of the body, vocalization, and drumming—the palm cockatoo advertises his territory.

THE MESSAGE IN VOCALIZATION

Many researchers still distinguish beween calls and song. The assumption behind this distinction has been that calls are short and simple and are produced by both sexes throughout the year, while song has at times been thought of as a special category of vocalizations reserved for male vocalizations during the breeding season. This distinction is no longer considered very useful, partly because of overlap (when does a call finish and a song begin?) and partly because not all song occurs only in the breeding season. In many songbirds, but by no means all, only the males sing and they are said to do so to attract a female. Other species, however, do not confine singing to the breeding season. In the tropics, many females sing. Also, in moderate climate zones, there are some species, such as the Australian magpie, in which males and females alike sing all year round. Some birds also have a song type that could easily be regarded as consisting of a few specialized calls.

From an evolutionary perspective it could be argued that patterns of vocalization may have become more common and more complex over time—that the most recently evolved species have the most complex vocalizations. The most recently evolved birds are the passerines, or songbirds, with about 56 families worldwide (from finches to scrub birds, swallows to pittas, starlings to flycatchers, pardalotes to crows, wrens to lyrebirds, warblers to currawongs—a very diverse group). Within this order we distinguish suboscines and oscines. Suboscines are birds supposedly equipped with a syringeal anatomy more primitive than that of "true" songbirds, the oscines. To the human observer complex song may be aesthetically more pleasing. However, suboscines' song may well have become more complex over evolutionary time. It would be easy to surmise that syntax or meaning is implied in the concept of complexity. But neither the complexity nor the beauty of the song is in itself an indicator of content. The actual communicative value of a long, beautiful, and complex song (such as that of the nightingale or of the lyrebird) may not be greater than that of shorter or less melodious vocalizations.

The frequencies of bird vocalizations commonly range between 2 and 10 kHz, frequencies that humans can easily hear and that are therefore easy to record and measure. Only a few avian species are known to produce infrasounds, such as the pigeon (sound levels down to 0.5 Hz), and a few species produce vocalizations in the ultrasonic range (above about 20 kHz). As noted in Chapter 2, the experimental technique of playback is the standard way of investigating the meaning of vocalizations. Playback involves recording the vocalizations of a bird and then playing them back to another bird or group of birds and observing the results.

Sending a vocal message can of course take many different forms, as we pointed out in earlier chapters. For birds (and many insects), which have such a high investment in communication by acoustic means, it is important to be aware that effective sending of messages may be impaired by factors in the environment in a number of ways.

We speak of auditory saturation, for instance, for sounds that are impossible to transmit over a long distance. Background noises such as wind, waves, rain, or the movement of leaves in a forest may cause wave reflections and bring about a lowering of signal intelligibility and a diminution in the carrying power of the signal. The background noise of other species, such as insects and frogs, may also interfere with transmission of the vocal signal. Then there is aggregate noise produced by the same species living communally in a colony, in large family groups, or in a bachelor flock, and even background noise created by the movement of wings. Hundreds of birds taking to the air at once can create a substantial noise even without vocalization. All these factors may impede communication of a message by sound.

The receiver must therefore be capable of extracting the relevant information from random background noise and have the capacity to detect information-carrying signals of an intensity even below that of the background noise. Who cannot be impressed when watching a penguin, for example, enter its colony of perhaps tens of thousands of other raucous birds and identify its young by sound alone. A message, at its most basic level, reveals the species identity of the sender. The study of birdsong makes it clear that each species has its species-specific vocalizations, although there may be individual variations, and even dialects according to region. Any vocalization therefore at the very least conveys the meaning: "I am here and I am a great tit" (or a starling, or a nightingale, or any other species).

We distinguish broadly between the syntax and the semantics of a message. Syntax refers to the structure of the song or call. Semantics refers to the content or meaning of the message. The two can be intertwined. It is conceivable that most vocalizations are intended to impart meaning. The exceptions are some vocalizations of a few cave-dwelling species such as the cave swiftlet *(Aerodramus vanikorensis),* which uses clicking sounds for echolocation just as bats do.

A bird's vocalization may be simple or complex. But it is often misleading to claim that a vocalization is "simple," because our fleeting observations of one species may not represent that species' entire repertoire and because we, as casual human observers, may not always be able to detect the finer distinctions in a vocalization. For instance, it is now known that the allegedly "simple song" of a finch has 13 themes and 187 variations. The entirety of vocalization variations in birds is called a repertoire, just as in a human singer, and the number of different song types available to one species is referred to as the repertoire size. Repertoire size has been examined in quite a number of songbirds. From available information, it seems that the brown thrasher holds the record in repertoire size, as Clive Catchpole and Peter Slater (1995) point out. The brown thrasher has an estimated repertoire size of between 1500 and 1900 song types. Improvisation and new learning may result in further changes and increases in the repertoire size. For instance, as John Kirn and his colleagues discovered in 1989, the red-winged blackbird adds to its repertoire each year.

But repertoire size, by itself, is not an indication of an increase in meaning. Meaning is far more difficult to assess than repertoire size. We use the term "vocabulary" to refer to the semantics—the actual meaning of the calls. Passerines may have a vocabulary of about 20 different calls, whereas gulls and other non-songbirds may have half that. However, ongoing research is constantly discovering more and more variation in avian vocabulary. Given that individual differences are very marked in vocalizations of complex songbirds, we may expect a good deal of variation and with that variation may come complexity of meaning.

The understanding of meaning in avian vocalization is in its infancy. Traditional ethology tended to describe animal behavior in terms of four main motivational systems: aggression, fear, feeding, and sex. These categories were related to physiological processes underlying the behavior. In 1953 Niko Tinbergen argued that behavior was due to relatively invariant and immediate responses to internal and external stimuli. This approach was an important first step in studying vocal and other behaviors systematically. Since his ground-breaking work, much research has been undertaken to investigate the development of vocal behavior in conjunction with physiological and even anatomical development. More recent studies have shown that vocal behavior in birds does not always conform to Tinbergen's simplified model. It is now known that learning plays a part in the development of song in all true songbirds so far studied. There is a period of vocal plasticity—a period during the development of the young bird when it is able to extend its vocabulary and learn its song. Even in those avian species with simple calls some learning may be involved. Fernando Nottebohm and his colleagues (1990) showed, for instance, that learning is enhanced or decreased by the acoustic context. Zebra finches that were asked to solve the problem of a missing harmonic in an experiment learned the operant response in a fraction of the time when the specific problem was embedded in a whole song (Nottebohm et al., 1990). The period for learning may vary widely between species. In some species of sparrow, learning is restricted to the first two months of life, while in others it may go on much longer. Peter Slater showed that the young chaffinch is able to learn new songs as late as 10 months after birth (Slater, 1989). The vocalizations of Australian magpies remain highly plastic throughout the first year of life at least. There is evidence from hand-raised magpies of learning of new sounds and new (human) words throughout this period (Kaplan, 1996). More details about learning to vocalize are given in Chapter 6. Here we want to emphasize that avian vocal behavior is extremely complex and certainly not automatic or based simply on underlying physiological factors.

WHAT IS SONG FOR?

The functions that have been established for birdsong can be summarized as territorial defense and sexual attraction. Donald Kroodsma (1996) has argued that sedentary species may develop elaborate songs, whereas migratory birds may use song to a lesser degree for the purpose of advertising their nesting or transient territories. During the breeding season, the growth of male sexual organs may be accompanied by changes in plumage (as in the superb blue fairy wren, for example) or by the onset of elaborate song for the purpose of attracting a female. Vocalizing to defend territory is generally regarded as a more efficient way of communicating than physical confrontation. Less energy is expended in the process and injuries may also be minimized. Many bird species first issue warning calls to an invader but then follow the calls by direct flight at the invading individual if the vocal warnings were not sufficient to deter the invader. Neighboring birds know their territorial borders, and a form of truce, even if a watchful one, may exist between neighbors. This is illustrated by the behavior of the white-throated sparrow *(Zonotrichia albicollis)*, which sings far less energetically when a neighboring bird approaches its territory than when a stranger approaches.

The connection between song and breeding is equally strong. There is ample evidence today that many males sing to attract a female just as some choose plumage to achieve the same result, and males of some species do both. Song requires energy, and one of the arguments put forward is that prolonged and strenuous singing advertises the good health and fitness of a male, just as a shiny and colorful plumage may. A study of the great reed warbler *(Acrocephalus arundinaceus)* has shown that females select males with larger song repertoires and that the survival of offspring is increased by such a choice (Hasselquist, Bensch, and von Schantz, 1996), and the same is known to be true of other species (Catchpole, Dittami, and Leisler, 1984). Apart from its possible physiological function across a variety of songbirds, the communicative value of the song may be: "Take me because I am healthy." Further, the song may say, "I am experienced and will therefore make a good partner." The singing male may convey an honest signal because an accomplished song is a mark of a mature adult, a bird that has had plenty of exposure to his species-specific calls. The male great reed warbler, for example, increases his repertoire with age and this expansion is a mark of his proven ability to survive.

Research done in the 1960s has shown that auditory stimulation has a direct effect on the secretion of hormones that stimulate growth of the sexual organs, which, in turn, stimulates the secretion of sex hormones. The secretion is induced by sound and, in some species, triggers the female to become ready for mating. A classic study by Daniel Lehrman (1965) showed that the cooing of the male dove triggers reproductive changes in the female.

Singing Together

Duetting has been an important subfield of song study and it occurs in a wide range of avian species. It is now recognized that duetting plays an important role in the vocal communication system of birds, especially those that live in the tropics. When talking about music sung by humans, we tend to use the term "duetting" for two voices singing not just simultaneously but in some agreed and orderly fashion that produces harmonies or contrasts in sound structure. Duetting in birds refers to the process of B starting to sing when A has stopped singing and A continuing where B left off. Calls made by duetting birds may overlap but usually the calls of two birds follow each other so closely and so precisely that they sound like the vocalizations of one bird, a phenomenon called antiphonal song (see Figure 4.3). In short, a duet is an agreed-upon sequence of calls that fit together owing to the choice of frequency, rhythm, and even overtones (harmonics). Duetting is now thought to be a specific form of communication, a way of retaining auditory contact especially in densely forested environments, which make maintenance of visual contact difficult.

Although duetting may play a part in synchronizing the gonadal state of the pair (to prepare for breeding), its functions also include communication when visual contact is lost or at risk of being lost. This is particularly true in wooded areas and dense rainforests (hence the prevalence of duetting in tropical regions) or during winter flocking and migration. Duetting may also serve to synchronize defense of a territory or, more commonly, to reinforce a pair bond. Duetting seems to occur more frequently in pairs with a prolonged monogamous bond. Australian magpie larks *(Grallina cyanoleuca)* duet regularly (Figure 4.3), as do Australian magpies, the black-faced cuckoo shrike *(Coracina novaehollandiae),* and the bar-headed goose *(Anser indicus),* but the contexts in which these species duet seem entirely different. Studies by Charles Blaich and his colleagues (1996) found that pair-bonded zebra finches engage in contact-call duets far more frequently than unpaired finches, and in a nonrandom fashion. Duets are not necessarily initiated by the male. In the bar-headed goose and the bay wren *(Thryothorus nigricapillus),* for instance, it is the female who calls first, answered by the male.

Further types of singing together are choruses and caroling. In the chorus, a whole group of birds sings at the same time, sometimes eliciting countersinging by neighboring and competing birds or by unrelated groups. There is a form of chorus that we call caroling, which occurs when members of a family group or communal breeders reconfirm their bond and, together, announce their possession of their territory. Austra-

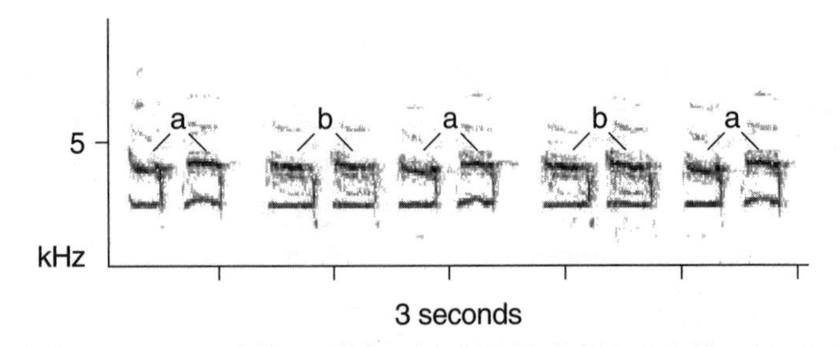

FIGURE 4.3 The antiphonal song of two Australian magpie larks *(Grallina cyanoleuca).* The vocalizations produced by the two individuals (a and b) as they duet are indicated. The marked similarity of the vocalizations by the two birds and the precision of timing make it appear as if the vocalizations were made by just one bird. (Sound spectrograms produced by G. Kaplan.)

lian magpies and kookaburras (also called laughing jackasses) use caroling and countersinging to test the strength of a neighboring group. As in individual calls, in caroling and chorus singing there may also be some status signaling involved. The parent bird starts to sing and then is joined by its mate and offspring or helpers at the nest. In kookaburras (laughing and blue-winged), the offspring may supply a form of percussion support while the parent birds burst into full staccato calls ("laughing").

Variations and Richness

The loudness of a vocalization (amplitude) can make a substantial difference to a message. Many bird species, as Richard Andrew has found, have loud-faint pairs of song display (Andrew, 1961). The loud vocalization may be for territorial display and can mean that the caller would attack if the territory borders were infringed. For instance, Carolina chickadees and Australian magpies have a vocalization that is uttered only when they are ready to attack. A fainter call signaled to mates and offspring may indicate that the communicator is ready to interact but not to attack.

Thus even if we consider only those vocalizations used to communicate messages about breeding and territoriality, the variations and richness of the message can be substantial. Some species of gulls even have specific copulation calls. Among weaverbirds (songbirds of the subfamily *Ploceinae*), John Crook found that, in at least four species, females have specific vocalizations they use to solicit copulation (Crook, 1969). In Australian magpies, females and males use the full song repertoire all year round. Bird vocalizations can also signal information about food; they can express anxiety or alarm, rivalry, interest, or defense readiness; they can tell others to fly away (follow me) and convey similar short instructions. None of these may be specific to the sex of a bird. Male and females alike will utter calls when predators approach and in many other situations.

Attributing specific functions to birdsong is useful in ascertaining the meaning of the vocalization and what evolutionary advantages might flow from one activity (song) over a host of possible others. But this approach may overlook or underplay other aspects of song. To consider song only in terms of territory defense may underplay how it is learned. Simply establishing a relationship between song type and territory may lead one to overlook, as J. M. Williams and Peter Slater (1990) have pointed out, that both repertoire size and numbers of neighbors are likely to have strong influences on the distribution of song types in a population. They conclude that geographical variation of song may be an epiphenomenon of vocal learning and that one need not propose purposes for geographical variation in song or for song dialect boundaries.

Straightforward functional explanations of song may also be unable to account for a whole range of other vocalizations. For instance, one of the magpies that Kaplan raised spent an average of 2 hours per day vocalizing, presenting elaborate variations of a long repertoire, for no apparent reason. The vocal pattern was rearranged every time and seemed to record the auditory events of the day, embellished perhaps, but recognizable as auditory events that had occurred within the bird's earshot. Over 20 percent of these vocalizations, recorded over an entire year, consisted of mimicry of sounds of sympatric species (those living in the same area)—kookaburras, peachface parrots, dogs, and humans. Free-ranging Australian magpies show the same behavior at the height of summer, well before their breeding season and just after the worst pressures of feeding their young from the previous season have abated and the food supply is plentiful. These examples provide some evidence to suggest that singing may increase as the pressures to defend territory and/or the young decrease (quite opposite to the claim that singing increases purely for the purposes of breeding and defense).

There may be "cultural" aspects involved in singing at this time. For instance, magpie females may sing to their offspring while feeding them. This vocalization may include not just the individual song of the mother but also mimicked sequences. In one recording made by Robert Carrick, Norman Robinson, and Bruce Falls in the mid-1960s in Canberra, a magpie mother "sang" something that sounded like a horse's neigh to her offspring just before feeding them (original tapes acquired by courtesy of Emeritus Professor Bruce Falls); and we now have a number of examples of similar mimicry across Australia (see below). These mimicked vocalizations might be cultural and have no direct survival function, unless it can be argued that conveying the information that horses were in the territory was vital knowledge for survival. Alternatively, this vocalization could have been a by-product of another form of mimicry that was vital for survival.

Meaning

Research on any aspect of semantics in the last 50 years or so differs from earlier studies in one exciting way. It is now known, and still being discovered in more bird species, that birds are capable of referring to objects outside themselves (called external referents) and can communicate this knowledge to others (through "referential signaling"). Studies in the 1950s, for instance, showed that some bird species signaled to conspecifics that they had found a particular food source. This was observed in herring gulls *(Larus argentatus)* by Hubert and Maple Frings (1956), and shown in a study by H. Friedmann of the African honeyguide *(Indicator indicator),* which leads conspecifics to nests of wild honey bees (Friedman, 1955).

The most common area of investigation of external referents concerns alarm calls. It has been shown for a number of bird species that the alarm calls for aerial predators are different from those for ground predators (see Chapter 2). In fact, Peter Marler (1981) noted some years ago that warning calls about aerial predators have similar acoustic qualities among very different species of birds. Whether uttered by a chaffinch, a blue tit, a blackbird, or a reed bunting (all European birds), the warning call is delivered with approximately similar intensity and at about the same pitch, 7 kHz.

The source of the warning call is not easily located. To explain this, we have to digress briefly into the physics of sound. Hearing and locating a source are usually achieved by both ears (a binaural function); the ears assess and compare crucial elements of the message such as phase, intensity, and time difference, thereby decoding the message and the location of the sender. Phase differences (the differences in time between when a sound wave reaches one ear and when it reaches the other) can be detected more effectively at low frequencies. At higher frequencies, the wavelength of sound decreases, rendering phase difference more difficult to detect, depending on the size of the listener's head, and hence the source more difficult to locate. In fact, depending on how far apart the listener's ears are (and therefore how large the listener's head is), there is in each individual case one frequency of sound whose source is impossible for the listener to detect by using the difference in time between the sound's arrival at each ear. If a bird used this frequency as the frequency of its alarm call, its predator would be completely unable to use timing to locate its potential prey.

Identifying the location of the sound source is further aided by the so-called sound-shadow intensity effect. If, for example, the sound source is to the listener's right, the left ear will be in the "sound shadow" of the listener's head. An intensity difference thereby occurs between ears, and this disparity can help establish the direction from which the sound comes. If a bird wants to avoid detection, it should pitch the call at a frequency that makes phase difference ambivalent and minimizes the sound-shadow intensity effect. By doing so, the bird can prevent clear identification of the direction of the call and hence can call to warn of the presence of a predator without running an immediate risk of being caught and eaten.

Peter Marler (1955, 1981) showed that a call of about 7 kHz does exactly that, and then showed that several species of birds use that frequency for alarm calling. This research suggests that certain sets of alarm calls may become common to many species because of their physical properties. Discovery of such rules of communication also make it more understandable why communication between very different species is possible. Marler found that alarm signals in some Corvidae and sparrows have similar structures and therefore induce interspecific reactions. In other words, the alarm call of one species may benefit a variety of other species, as we saw in Chapter 2.

Many other signals used by avian species have not been fully investigated, but some of them are certainly known to pet owners and those who rehabilitate wild animals. Among them are signals that indicate emotions. For instance, dogs have been known to cry for their owners. Birds shake in fear while they utter species-specific (often barely audible) high-frequency vocalizations. Animals communicate their emotions and desires to humans, and pets have also been observed to communicate with other species. Robert Leslie (1985) described a case of deception based on interspecies communication between two birds, a parakeet and a blue jay. The visiting parakeet, perched on the outside of the jay's cage, seemingly hungry, indicated by eye position and other cues that it wanted the chopped spinach in the cage. The blue jay moved the chopped spinach close to the edge of its cage, but on the inside, and when the parakeet reached for the spinach the blue jay attacked the parakeet's head with its beak.

Attachment may be expressed by a combination of preening behavior and low gutteral sounds. For instance, low-frequency gutteral sounds are emitted by Australian magpies when they preen a partner or offspring or the human who cares for them. Galahs emit a specific short "approval" call in conjunction with the response of another galah (or human), and during preening the preened bird purrs much like a cat.

It is interesting to note here that similar patterns of intonation occur across human cultures. Anne Fernald, for instance, has shown that in humans melodious speaking signals approval; sharp, staccato bursts express disapproval or denial; and low legato murmurs are meant to comfort—and these patterns are common to different cultures (Fernald, 1992). These patterns may apply not only to humans but to animal species as well. Sharp calls are usually interpreted as repudiating calls while low legato murmurs/purrs are associated with comforting: there is cross-species similarity, as Marler (1981) described for alarm calling.

Mimicry

Birds have another set of vocalizations not equaled by any other group in the animal world: mimicry. Mimicry is extremely widespread and highly developed among Australian bird species but is found also throughout the rest of the world. The Australian species best known for mimicry in the wild are both species of lyrebirds (Robinson, 1991), Australian magpies, and bowerbirds (several species). In contact with humans, even if remaining free, they can also mimic human speech. Among European birds, the starling is the star of mimicry (Hausberger, Jenkins, and Keene, 1991). We know that parrots and budgerigars are excellent mimics in captivity, but the first examples of mimicry in the wild have been found only recently, for example in the Grey parrot (Cruickshank, Gautier, and Chappuis, 1993). European marsh warblers *(Acrocephalus palustris)* copy the calls of over 70 different species that they hear in both Africa and Europe, between which continents they migrate (Dowsett-Lemaire, 1979). It is from such mimicry that young birds of this species are thought to learn their songs; they cannot learn them from their fathers because their fathers cease to sing before the chicks hatch.

The question remains: what is mimicry for? Why would birds deliberately transgress their species-specific sounds and move into the vocal territory of other species? We know that insects can mimic appearance, smells, and even noxious taste signals, and that dolphins and seals may use some vocal mimicry, but as far as we know today only birds mimic other species extensively in their vocalizations. Purists argue that such mimicry by birds is not "true" mimicry; they define "true" mimicry as having deceptive purposes useful for survival.

According to the models derived from studies of the insect world, true mimicry involves three parties: the mimicked one, say butterfly A, the mimicker, called butterfly B, and the predator that is fooled by butterfly B—the predator will not eat B because it looks like the unpalatable butterfly A. There has been no unambiguous evidence to date that birds mimic to avoid predation. However, it is possible that a bird may mimic another to safeguard a territory. Although this is not mimicry to avoid predation, it clearly functions to aid survival, either by safeguarding a territory from a rival or by repelling a predator who may prey on the young in the nest. In considering mimicry, we must also take into account differences between intentional and unintentional vocalizations, as we did in Chapter 3 for signaling in general (see the 1997 review by Christopher Evans for more detailed discussion).

A second reason for mimicry, and the one most commonly cited, has to do with the breeding season. Lyrebirds, for instance, adorn their songs during the breeding season with all sorts of sounds, taken from the sound repertoire available to the male. These added sounds typically include mimicry of the sounds made by other birds, the most distinctive being currawongs, kookaburras, yellow-tailed black cockatoos, and catbirds (mostly species that mimic others themselves). Lyrebirds may also include sounds of barking dogs, car horns, creaking door hinges, and even chainsaws (Robinson and Curtis, 1996). Male lyrebirds sing their long sequences of mimicked calls to attract a female. It is as if they "wear" the song component like medals—the more elaborate and extensive the collection, the more the female may be impressed.

But the function of vocal mimicry may extend even further, at least in the case of the Australian magpie. Mimicry in magpies has now been recorded from all over Australia, and preliminary data have shown that sounds are mimicked very selectively (Kaplan, 1996, 1999). Extensive exposure to some sounds resulted in no mimicry, while very short exposure to others immediately produced mimicry. The conclusions drawn so far are that mimicry occurred only in species that shared the same territory. Visitors, transient species, seemed systematically excluded.

In this case it seems possible to argue that territorial knowledge is very important for a species that is highly territorial, and is incorporated into the magpie's own repertoire. The sonogram in Figure 4.4 shows that the mimicry of the kookaburra is more melodious than the original, but the

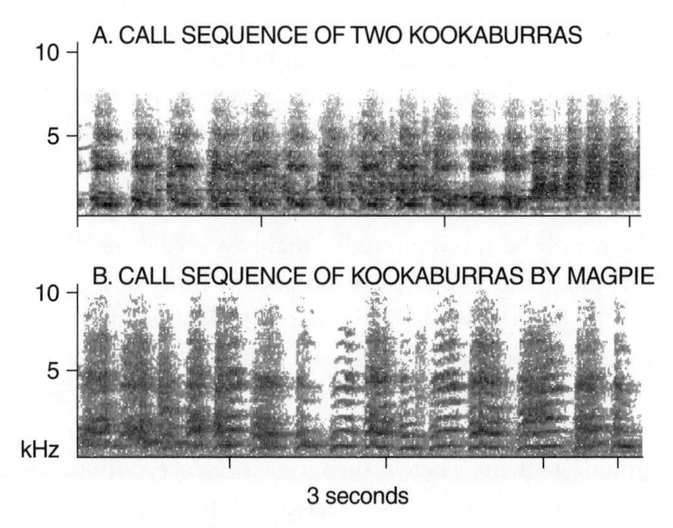

FIGURE 4.4 Mimicry of a kookaburra by a magpie. A: A sound spectrogram of two kookaburras *(Dacelo gigas)* laughing jointly. B: Mimicry of the kookaburra's laughing by an Australian magpie *(Gymnorhina tibicen).* Note the matching rhythmic pattern. The fundamental notes match also but most of the overtones do not. The magpie's rendition of "kookaburra" is rhythmically very precise, but it is a little more melodious than the original, demonstrating the vocal differences between songbirds (passerines) and non-songbirds (nonpasserines).

magpie has attempted to follow the rhythmic patterns of the "laugh" rather precisely, mimicking not just one kookaburra but the joint calls of two birds.

The quality and complexity of bird vocalizations has also raised the issue of whether some bird species may be capable of communication that approximates aspects of language. First, it needs to be noted that "communication" and "language" are two different concepts. We have already seen that effective communication is possible by means other than sounds and language. Second, there is the issue of how we define language. Well-known contemporary linguists such as Steven Pinker (*The Language Instinct,* 1994) and Derek Bickerton (*Language and Species* 1990) have made a strong case for the species-specificity of human language, arguing that human languages are qualitatively different in structure from systems of animal communication. We have no problem with defining human language as "species-specific"—saying that human language has attributes found exclusively in humans—since other, nonhuman, species also have unique attributes. However, what is occasionally asserted, and often only implied, is not just the uniqueness of human language but the uniqueness of the processes required to achieve language—that is, intelligence.

There have now been several studies that have challenged the view that language is unique to humans. For instance, a study of Japanese quails undertaken by Keith Kluender and others (1987) showed that quails can learn phonetic categories. These results challenge theories of speech sounds that posit uniquely human capacities. Irene Pepperberg has demonstrated that her parrot Alex understands commands and concepts and can communicate them. There is then some evidence, both phonetic and semantic, that not all processes associated with the acquisition of human language are unique to humans.

CONCLUSION

Avian species have certainly developed great virtuosity in both vocal and visual communication. For this reason alone human fascination with birds will continue to be strong. It is clear from research so far that some, if not all, of the signals made by birds are not merely emitted reflexively, but involve learning and quite complex decision making, depending on the social context. Some of the astounding vocal abilities that certain bird species share with mammals are almost certainly the outcome of convergent evolution, meaning that these abilities evolved separately in the avian and mammalian lines of evolution. Other special features of bird communication probably existed in vertebrates long before mammals evolved and used them. But whatever their evolutionary origins, the vocalizations of birds follow principles or rules that are relevant to other species. As Nosumu Saito and Masao Maekawa (1993) and many other researchers have pointed out, comparing avian vocal communication with human vocal communication can be most instructive.

Chapter Five

COMMUNICATION IN MAMMALS

There has been less well-controlled experimental research on the communication systems of mammals than on those of birds. Most recent research on mammals has found that their communication systems are more complex than was once thought, and that many of these signals vary according to the context. As in other species (see Chapter 3), some signals are sent unintentionally and reveal something of which the sender is not aware, whereas others appear to be sent intentionally.

VISUAL MESSAGES

The visual signals used by mammals are diverse and complex, but there are basic signals that strongly resemble each other across many mammalian species, including humans, and some of them appear to have been used over long stretches of evolutionary time. Visual signaling in mammals is usually confined to changes in body posture, such as stretching, jumping, arching the back, and moving the limbs. Tool use purely for visual signaling does occur in mammals, but it is relatively uncommon.

In our discussion of the male bowerbird's use of his bower in courtship displays, we noted that visual displays may involve objects other than the body. Such use of objects in visual displays is of importance in avian communication, and there are even examples of fish, amphibian, and reptilian species using objects for display. It is therefore significant that mammals do not often use objects to enhance visual displays. One of the few examples cited in the literature is object use by wild orangutans, observed by John MacKinnon (1974). When MacKinnon kept following them through the rainforest, they looked down from the trees and started to throw sticks. Some of these "weapons" barely missed him, and in a few cases stick throwing became more intense when he continued to follow the orangutans. MacKinnon rightly read this behavior as a warning signal that he should stay away. Another exception may be the gibbon's branch shaking, which some researchers, such as Peter Marler and Richard Tenaza (1977), have regarded as a visual rather than an auditory signal. We also know that baboons throw stones at predators.

Visual displays may be categorized according to which part of the body is used. The body as a whole, including its posture and movement, can function as a signal. Locomotion itself is a form of communication. A particular gait and the corresponding body postures may well determine how another animal will respond. Limb movement is a separate aspect of visual display. For mammalian species with tails, the tail may be used extensively to accentuate the meaning of the animal's emotions or intent, or it may even constitute a signal on its own. In many ungulates (hoofed mammals) and carnivores the tail is used in greeting, threats, and courtship, each with its characteristic postures and speed of movement. Monkeys also use the tail extensively in friendly and aggressive displays (see Figure 5.1, which includes tail and genital display), as Richard Andrew (1972) has shown. In Chapter 2 we discussed the use of tails by lemurs in "stink fights": raising of the tails and waving them so that odor is wafted toward the other animals. We did not mention the visual aspects of this display: the long tail of *Lemur catta* is dramatically striped in black and white. In moonlight this striping would be visible and it may constitute part of the signal. In many species, the movement of the tail mirrors the movement of the head in several displays. For instance, holding the head and tail high signals high arousal and/or dominance. Lowering of the head and tail signals submission and even fear. Familiar body movements in our own human signaling system are shared with many other mammalian species—waving, head shaking, jumping, raising the arms, and so on.

Another region of the body from which signals can emanate is the face (Figures 5.2 and 5.3). Most of the research on faces has been carried out on primates and this is a very large field of investigation. Facial expressions and nonverbal communication in primates have been of interest partly because a primate's face is similar in anatomy to the human face and partly because Darwin singled out the face as an important site for the expression of emotions (see Chevalier-Skolnikoff, 1973, or Ekman, 1974, for details).

Early work by Jan van Hooff (1967) showed that there are possible primate homologues of both human laughter and smiling. The nearly silent bared-teeth display has been described as phylogenetically one of the oldest facial expressions, shared as it is not just by primates and humans but by many other mammals as well. Usually this gesture is associated with a threat or strongly aversive stimulation. A silent bared-teeth display is a sign of fear and submission, found in many higher primates. Our human smiling may have arisen from this facial expression, although, if so, it would have had to undergo a change in meaning from its agonistic origin to become an expression of attachment. Or the human smile could have

FIGURE 5.1 Genital display of a marmoset. The marmoset on the right is displaying its genital region to the human taking the photograph. The marmoset on the left is looking at the display and is probably receiving an odor released from scent glands in the anogenital region of the displaying marmoset. (Photograph by the University of New England Media Unit.)

its origins in the play face (see Figure 2.4), and some consider this more likely. Van Hooff believes that human laughter and human smiling have different phylogenetic roots; he thinks laughter arose from displays of fearful submission. Laughter is associated with breathing and breathing technique—that is, a vocal activity—but smiling is solely a movement of facial muscles. Chimpanzee laughter is closely coupled with breathing, but Robert Provine (1996a) found that unlike humans, who exhale continuously when laughing, chimpanzees produce one laugh sound per expiration and inspiration. Signe Preuschoft's study of Barbary macaques *(Macaca sylvanus)* further confirmed the phylogenetic difference between the smile and laughter made by van Hooff 25 years earlier (Preuschoft, 1992).

Another facial expression found in many mammals as well as in humans is the yawn. Yawning is probably more widespread than the descriptions in the literature indicate. When humans yawn, they are usually indicating that they are tired or bored, as a detailed study by Robert Provine confirmed (Provine, 1996b). Yawning is thought to be "contagious," like laughter, and a yawn can also be sent as a signal to express disapproval. In mammals, yawning may mean a variety of things. In baboon parlance, a yawn by itself either signals uncertainty or expresses fear; in the latter case, the yawn can become part of a signal of aggression when other body movements are added. By contrast, dogs may yawn when they have been praised, or yawning may express frustration.

Other facial expressions that we share with primates are grimacing, tongue movement, staring, certain eye movements, and expressions of sadness. The eyes play an especially large role in facial expressions that humans share with primates. Often in concert with other facial expressions, the eyes can express fearfulness, anger, curiosity, and real or feigned indifference. The stare, as Jean-Pierre and Anne Gautier (1977) have pointed out, conveys several meanings for Old World monkeys. One is a threat. Another is a reprimand. A male gorilla will use a stare if grunts are unsuccessful in settling squabbles between females or juveniles. In dogs, a stare can also express a wish or demand. Dogs often stare when they beg from humans or each other.

Juichi Yamagiwa (1992) pointed out that the role of stares in primates is rather complex and may have different context functions among bonobos than among rhesus macaques or gorillas. Bonobos and chimpanzees may use mutual staring as a form of positive contact with each other, while the stare of the gorilla without any subsequent physical contact may be used in conflict resolutions. In the late 1950s Niko Tinbergen suggested that displays of any kind can be conveniently divided into those that are distance-increasing and those that are distance-decreasing. Peter Marler (1968) extended this classification by suggesting that pri-

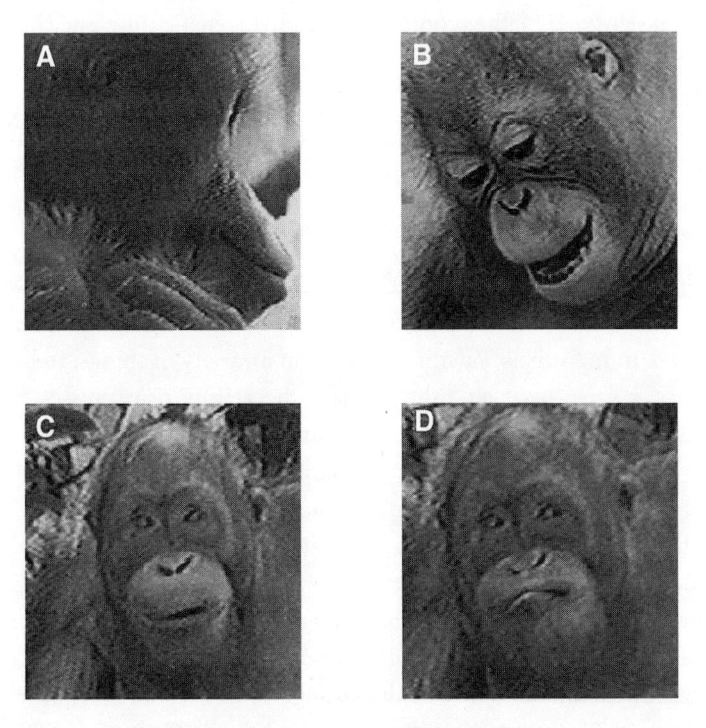

FIGURE 5.2 Facial displays of orangutans. These are Bornean orangutans *(Pongo pygmaeus)*, which we filmed in Sabah, East Malaysia. A: Jessica is holding her 1-week-old baby and smiling, just as a human might. B: A play face. The lips are puffed, the mouth is partly open, and the lower teeth are just barely showing (compare the play-threat display in Figure 2.4, where the mouth is opened wide and both the upper and lower teeth are displayed). C and D: Two frames from a videotape taken in close succession. This young male is expressing mild anger first by parting the lips to grunt (C) and then by protruding the lips into a pout (D). (Videotapes by G. Kaplan.)

mate communication functions to achieve either aggregation or dispersal. Moreover, as has been observed in bonobos, a direct stare may be a means of seeking a sexual encounter or it may be a reprimand or an assertion of dominance.

Eye stares in primates are often accompanied by lowering of the eyelids. The eyelids then become exposed, and in some primates these can be quite spectacular. In various macaques they are white and in some baboon species a silvery color. If the eyelids remain exposed or are rapidly opened and closed, the signal is made more threatening or at least more conspicuous. The human habit of painting the eyelids for accentuation is an interesting custom in view of primate signaling with the eyelids.

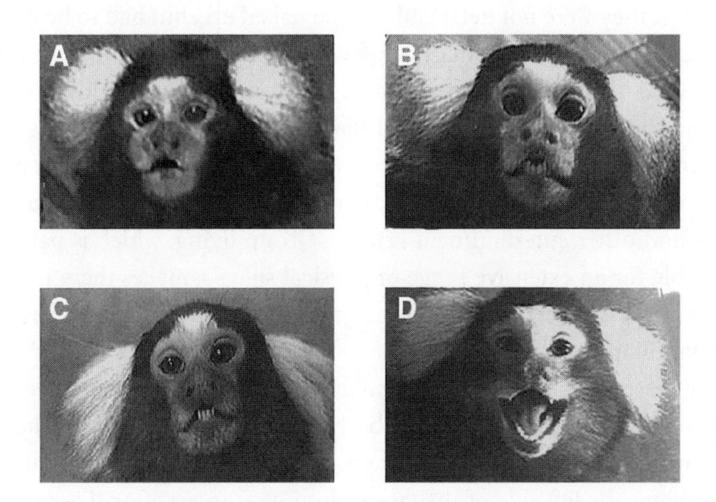

FIGURE 5.3 Facial displays of marmosets. These images of the common marmoset *(Callithrix jacchus)* are taken from videotapes. A: The facial expression that accompanies the twitter call, given to initiate social contact (see Figure 5.4). B: A face expressing apprehension or mild fear and threat. Note that the mouth is drawn back and the lower front teeth are displayed. When a human makes a facial expression like this in the presence of marmosets, they become very agitated. C: Expression of a level of fear higher than that shown in B. The mouth is similar to that in B but the ear tufts are lowered. D: The highest level of fear/threat is expressed and the marmoset is making the mobbing call, a rapidly repeated tsik sound). The mouth is open wide to display all the lower teeth and the incisors of the upper jaw. The ear tufts are pulled back. (Video images courtesy of M. Hook-Costigan.)

Raising the eyebrows either by retraction of the scalp or independent eyebrow movement further reveals the eyelids. Jean-Pierre and Anne Gautier (1977) point out that in various mangabeys and baboon species scalp movement is accentuated by side whiskers or the raising or flattening of tufts of hair on the top of the head.

There is also a furtive expression used by humans and orangutans when they want to look at something they ought not to. An extension of this kind of eye movement is the flirt. Our own work on eye movements in orangutans has shown that eye movement is employed more often than head movement. We have also observed "flirting" orangutans who edged closer and closer to each other and only occasionally looked at each other from the corner of their eyes. These eye movements were so brief that they were not detectable by the naked eye, but had to be discovered in frame-by-frame analysis of videos (Kaplan and Rogers, 1996). Additional signals that humans, as a species, have largely lost, such as lip smacking, ear flattening, eyelid flashing, and hair bristling, are common in most primates. Tragically for monkeys and apes, their very physical expressiveness has made them desired "objects" for display in circuses, clubs, and other entertainment centers. Group living, which is partly responsible for an extensive range of physical signals, makes them interesting animals to observe, and their signals, even if often misunderstood, look familiar to the human species.

The reason for the perceived similarity among humans, monkeys, and apes of visual signals given by the body and especially by the face lies not only in the morphology of the face but in associated brain mechanisms. In humans, the left side of the face is dominant in emotional expression. In Marc Hauser's study of the facial expressions of rhesus monkeys and the human response to their expressions, he found that facial expressions in rhesus monkeys begin earlier on the left side of the face and involve larger movements of the facial features than on the right side of the face (Hauser, 1993). Thus the left side of the face is more expressive. This has to do with the control of such expressions by the brain. The right hemisphere of the brain controls the left side of the face and the right hemisphere is involved with emotional expression in a range of species. By contrast, the left hemisphere of primates and other species processes species-specific calls. In fact, human and nonhuman primates have the same pattern of brain asymmetry for sending and receiving vocal signals. Alan Fridlund (1994) has recently warned, however, that we should not jump to the conclusion that commonalities necessarily mean a shared genetic heritage. Nevertheless, since other mammals and even birds and amphibians appear to have a similar asymmetry, it now seems likely that these characteristics do reflect a shared genetic heritage (Bradshaw and Rogers, 1993).

SOUND SIGNALS

Mammals frequently use sound for communication, sometimes sound within the hearing range of humans but also outside it. The range of frequencies used outside human hearing may be below (infrasound) or above (ultrasound) the thresholds of human auditory perception.

Echolocation

Use of ultrasound by animals was discovered in the twentieth century, largely through the work of H. Hartridge and G. W. Pierce. Hartridge worked on bats and concluded that they were able to avoid objects in flight by listening to the echo of their own sounds in the ultrasonic range. Research on echolocation was expanded significantly by the work of Donald Griffin (see Griffin, 1958). Pierce took up entomology as a hobby and later wrote about ultrasonic sound in crickets.

This discovery of the use of ultrasonic sound for navigation and communication was more important than we might think today. It opened our minds to the possibility that our own senses may not suffice for a full understanding of animal communication. Human audition ranges from about 0.02 kHz (20 cycles per second) to a maximum of about 20 kHz (20,000 cycles per second). The most sensitive and comfortable hearing for humans lies at frequencies around 2 kHz. As we have seen already, this frequency range is also commonly used by birds. We now know that, as well as a range of insects—from moths to grasshoppers, crickets, and locusts—there are rodents, whales, dolphins, seals, sea lions, and certain primates whose vocalizing and hearing range extends well above that of the human species. John Altringham (1996) showed that some bats of the Megachiroptera family use echolocation and that all bats from the suborder Microchiroptera, which includes hundreds of species in Old and New World areas, use echolocation, whether they are omnivorous, insectivorous, or carnivorous.

Some species of bats use echolocation not just to detect objects (to be avoided in flight) or potential prey but also to locate conspecifics. For instance, between birth and weaning the pups of the Mexican free-tailed bat *(Tadarida brasiliensis mexicana)* live in segregated colonies, or crèches, of about 4,000 pups per square meter. Each female has a single offspring and needs to locate it, usually twice in a 24-hour period. Gary McCracken (1993) established that a mother finds her pup largely by locational cues derived from echolocation.

The discovery of echolocation in sea-dwelling mammals, such as dolphins and whales (cetaceans), was made as late as the 1950s (by A. F. McBride for one) and later reported by Winthrop Kellogg (Kellogg, 1961). These two researchers found that the echo-ranging signals (clicks) are highly directional and extremely varied. The sonar characteristics of sea mammals differ, and each species has its own structures and frequency ranges. Transient killer whales, for instance, use short and irregular echolocation "trains" composed of clicks that appear to be structurally variable and low in intensity. Even within the same species of killer whales (*Orcinus orca*), there are substantial differences in the echolocation pulses. Whales that are resident in a region use regular sequences, while transient killer whales employ irregular sequences. As Lance Barrett-Lennard (1996) and his colleagues argue, sequences of short-duration sounds that are irregular in timing and frequency more closely resemble random noise than do sequences of more structured sounds and thus are less likely to be detected by marine mammals (the prey of killer whales) against background noise. By using these "noisier" sounds to locate their prey, transient killer whales can detect and approach marine mammals without their knowledge.

Signals as reliable as those used in echolocation can also serve a communicatory function. In the Microchiroptera there is evidence for a continuum between the use of ultrasound for echolocation and for communication. It has been shown by Brock Fenton (1994) and many others that the echolocation calls of one individual can be used simultaneously by other animals, both conspecifics and other species. For example, "feeding buzzes," which are echolocation calls with high repetition rates produced by bats when they attack airborne targets, indicate that prey is available and are often exploited by conspecific bats to identify vulnerable prey. Some moths also use "feeding buzzes" to detect the presence of predatory bats.

Yet echolocation does not necessarily give the killer whale a substantial advantage in catching prey, as Lance Barrett-Lennard and his colleagues found (1996). Although most fish species have auditory sensitivity in the low-frequency range of around 3 kHz and are thus unlikely to detect killer whale clicks, the typical prey of the killer whales can in fact hear the clicks. The pinnipeds (seals) and cetaceans (dolphins, whales) that are the prey of transient killer whales have acute hearing up to frequencies beyond 30 kHz, well within the range of killer whale sonar clicks. Porpoises swim away from killer whales at high speeds on erratic courses. Dolphins and gray whales move into shallow water when killer whales are nearby. Some prey have also adapted by emitting their own echolocation signals outside the range of killer whale hearing. For instance, killer whales cannot hear the echolocation pulses of Dall's porpoises, which center on frequencies of 135–149 kHz, because killer whales can sense frequencies only up to about 105 kHz. However, killer whales can hear the sounds generated when porpoises surface and breathe, and so they may find these prey animals without the aid of echolocation signals. Transient killer whales often search for prey in waters close to the shore, where there is camouflaging noise from waves striking the shore.

From Whistles to Roars

Other sounds that sea mammals make cannot be catalogued here—there are too many. Like those of birds, their vocalizations are species-specific, and often each individual has a characteristic pattern of vocalization. In the last 40 years there has been a great expansion of our knowledge of vocalizations in sea mammals. By the 1960s, considerable knowledge of the complexities of vocalizations of sea mammals had been acquired, and Roger Payne had described the vocalizations of humpback whales as song; there was even a record called *The Songs of the Humpback Whales.*

We now know that dolphins, for example, emit richly diversified whistles, usually at low frequencies. And we also know that their vocalizations, like those of birds, contain acoustic signatures of individuals and many sounds with precise meaning. M. C. and D. K. Caldwell (1965) first reported that bottle-nosed dolphins *(Tursiops truncatus)* had individually specific signature whistles. And a study by Vincent Janik, Guido Dehnhardt, and Dietmar Todt (1994) found that, beyond individual identities, the whistles also contain context-related information. Like birds, dolphins give different alarm calls in response to different predatory species, such as sharks, human beings, and killer whales. Perhaps the most spectacular dolphin behavior is the response to distress whistles made by other dolphins. A distress call by a sick or injured dolphin will bring other dolphins to its rescue. William Stebbins (1983) has reported that a dolphin having difficulty in rising to the surface to breathe air will be assisted by other dolphins and literally lifted up to the surface. Agonistic behaviors are expressed by jaw clapping or arching of the back. The tail flukes, the flippers, and the tail itself may be employed in such displays, and when accompanied by rising and falling whistles, they may indicate strong threats.

Seals have long been used in circuses because of their playfulness and ability to learn tricks. Seals, sea lions, and walruses (all pinnipeds) were, until recently, thought to command only a very limited range of vocalizations and to produce them only on land. It has been known since the 1980s that seals may mimic, but they usually do so when they are on land. A study in 1984 by Evelyn Hanggi and Ronald Schusterman of harbor seals *(Phoca vitulina)* found that they also vocalize under water during the breeding season. The vocalizations are different for each individual. Some vocalizations produced during the breeding season, such as the roar, are combined with visual aquatic displays. It has been suggested that such vocalization either plays a part in male-male competition or is a way of attracting females. Male walruses produce bell-like sounds that they use in combination with visual displays to attract females. However, it is perhaps a little premature to draw conclusions on the function and meaning of these display behaviors and the roars of seals.

Reproductive Strategies

During different seasons of the year, in accord with the reproductive cycle, mammals use different classes of vocalizations specific to communication about mating readiness and breeding. Many mammalian species have developed elaborate strategies surrounding the time of reproduction. Vocal, visual, or chemical signaling during the ovulatory period, the period of breeding readiness of the female, occurs in most mammalian species.

These signals function in a social context where it might be prudent for the female to advertise her reproductive condition or, alternatively, to hide the onset of the ovulatory period. Females may exploit their readiness for mating to entice males to fight on their behalf, or to ensure that they mate with the best possible male. In elephant seals *(Mirounga angustirostris)* the female gives a copulation call that incites aggressive competition between males. She witnesses the fight and the winner mates with her.

The mating strategies used in baboon societies are very different. The threat of infanticide is very real in baboon troops, and it occurs particularly in troops where only one male is present. An outsider male who successfully challenges the position of the troop male will attempt to kill offspring that are not his. Groups with several males are thought to be safer, partly because of the mating strategies employed by the female. The baboon female calls during mating, and as Sanjida O'Connell and Guy Cowlishaw suggested (1994), these calls may invite several males to mate with her, thereby creating uncertainty about paternity. This uncertainty might well protect her offspring. Baboon females also signal their estrous period by the reddened skin of the buttocks, which they display to males.

We know from other mammals that vocalizations may stimulate ovulation by the female. For instance, the roars of male red deer *(Cervus elaphus)* are said to trigger copulation readiness, if not ovulation, in females. Red deer males roar loudly and repeatedly during the breeding season (Clutton-Brock and Albon, 1979). Karen McComb (1991) found that the roaring rate was positively associated with reproductive success and fighting ability. The intensity, duration, and rate of the call may serve to advertise fitness in males. Although roars often precede fights with competing males, McComb found that male deer roar in the same way whether competitors are present or not and go on roaring at a rate of two roars per minute throughout a 24-hour period. These vocal displays are accompanied by other displays, such as waving of horns and broadside body movements showing off physical attributes. She found that females preferred males with a high roaring rate whether the males fought or not.

Primate Vocalizations

After the diverse group of mammals using ultrasonic sound (sea mammals and bats), the largest group of mammals in which vocalization has been examined and analyzed in great detail is the primates. Humans and most anthropoid primates have sacrificed high-frequency sensitivity for improved auditory discrimination within a restricted frequency range. In this lower frequency range, they have finer discriminatory powers in all three parameters of vocalization: frequency, intensity, and timing (temporal disparity). In addition, some mammals—marmosets and tamarins—are known to hear ultrasonic frequencies as well as frequencies audible to humans. Figure 5.4 shows vocalizations of marmosets within the human auditory range.

The fact that most primates are gregarious has led researchers to argue that the communication systems of such species are relatively complex, in accord with the complexity of their social organization. The Costa Rican squirrel monkey *(Saimiri oerstedi)* has become the most common laboratory model for studies of primate vocalizations over the past 20 years, especially calls expressing emotions and isolation. Most of these studies have been conducted in the artificial setting of a laboratory, supplemented only occasionally by studies in the natural environment. Laboratory work is important for controlled studies, but field work is needed to confirm what is found in the laboratory. One of the few field studies was undertaken by Sue Boinski in 1991. She found that, in the squirrel monkeys' natural environment, the duration of peep vocalizations (contact calls) is positively correlated with spatial separation, confirming that the duration of their peep calls provides information about the distance of the caller.

Some of the long-range vocalizations of primates are specialized calls that primates use when they discover food. Call characteristics are influenced by the quality of the food, its quantity, and its divisibility, as is also the case in birds. Marc Hauser (1993) and his colleagues found that chimpanzees emit a vocalization, which the researchers called a "rough grunt," when they find large amounts of food. Such vocalizations show

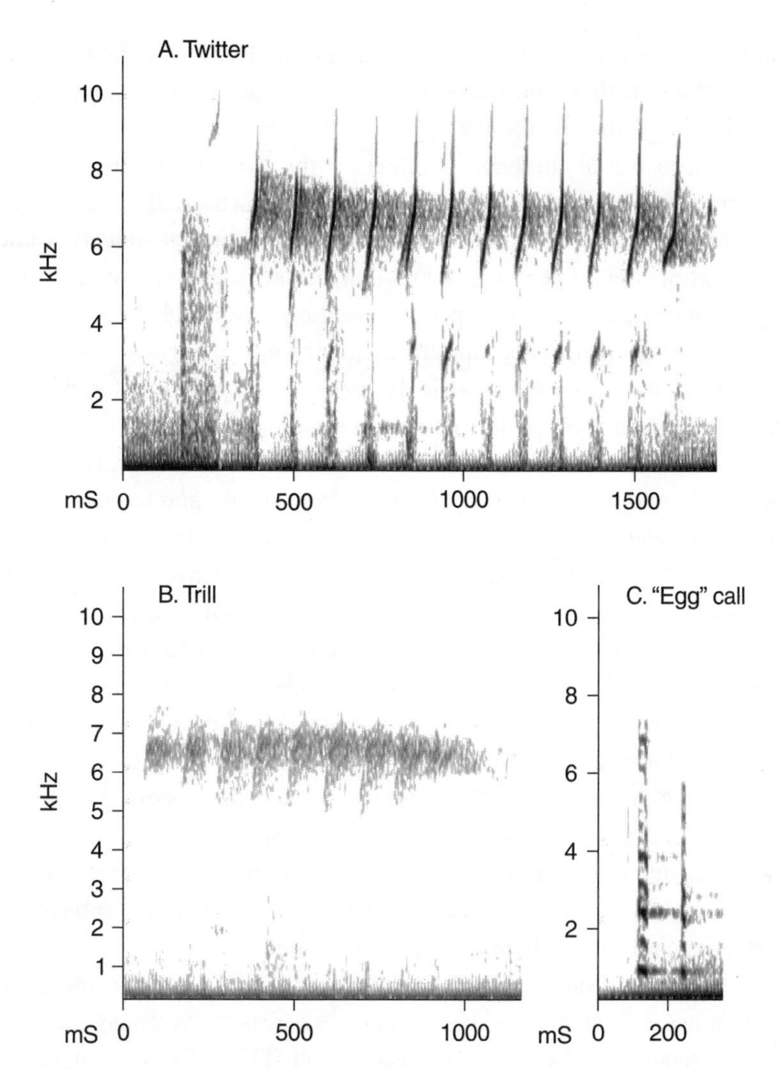

FIGURE 5.4 Vocalizations of marmosets. These calls were made by captive common marmosets *(Callithrix jacchus)* in our colony at the University of New England in New South Wales, Australia. A: Twitter calls, which are given to initiate social contact (see Figure 5.3 for the accompanying facial expression). Note the steep change in frequency. These calls extend into the ultrasonic range but the high frequencies are not represented in the figure. B: The trill call, given when the marmoset is slightly aroused. C: The crackle or "egg" call, which indicates mild alarm. These are just some of the vocalizations produced by marmosets. For more details see Epple, 1968. (Recordings by M. Hook-Costigan.)

that an individual is capable of making several decisions before vocalizing, such as "Is there more food than I need?" and "Can this food be divided among others without a fight?"

The acoustics of the habitat influences the structure of vocal signals by primates, as it does in all species. Charles Brown (1995) and his colleagues found that the rainforest is less favorable for high-fidelity sound propagation than open spaces but that, nevertheless, species that live in the rainforest have developed vocalizations with high-fidelity transmission by using the appropriate frequency and other acoustic qualities. Tropical rainforests, in particular, pose several problems for effective communication that do not occcur in savannas and open woodlands. The forest is an environment with high background noise, largely caused by insects, and high reflection of sounds from trunks and leaves. Foliage, temperature gradients, and ground effects can also contribute to fast degradation of the structure of a signal. Moreover, visual methods of long-distance communication are not readily available (trees and leaves also obstruct visual contact), and so forest-dwelling species have to use vocalizations for contact calling. Brown and his colleagues have shown that selection for vocalizations with a reduced chance of distortion has influenced the form of the vocal repertoire of two rainforest species (the blue monkeys, *Cercopithecus mitis,* and gray-cheeked mangabeys, *Cercocebus albigena*) more strongly than those of two savanna species (vervet monkeys, *Cercopithecus aethiops,* and yellow baboons, *Papio cynocephalus*). The forest environment leads to adaptations that overcome problems of distortion and thus influence the form of signal repertoires.

In rainforests individual primates lose sight of each other and must rely on acoustic signals to stay in touch, whereas on savannas they can usually retain visual contact with each other. This difference might explain other aspects of the evolution of vocal repertoires with different physical characteristics in primates living in rainforests and on savannas. Mangabeys and baboons vocalize in choruses, just like birds, enlisting the participation of many individuals. They scream, and although their vocalizations are very similar, the distortion scores are lower for the forest-dwelling mangabeys than for the savanna-based baboons.

The habitat of the New World squirrel monkeys is also densely foliaged, so in this species, too, there is a real possibility of the separation of members of the troop. Boinski (1991) noted that the female vocal exchange among squirrel monkeys acts as an "auditory beacon" to monitor the position of females and hence of the troop. To verify this, common marmosets were experimentally tested for responses to loss of visual contact. An experiment that deprived common marmosets (also arboreal monkeys of the South American rainforests) of visual contact showed that they immediately modulated their calls in duration, peak frequency, frequency range, and median frequency. Lars Schrader and Dietmar Todt (1993) found that modulation increased with decreasing sensory information about mates. Also, the amplitude of the mammals' calls increased as much as 8 decibels during the experiment, an escalation that improves transmission of the calls. Schrader and Todt concluded that modification of specific call parameters can protect information encoded in the calls against possible signal disturbances caused by the environment. Thus call modulation is found to be linked to spacing.

Most of the auditory signals of the forest species studied so far have been classified as messages related to the integration of the whole group, such as alarm calls or contact solicitations, but such calls, of course, are not limited to forest dwellers. Alarm calls may even be specific enought to relay information about the type of threat that is imminent—signal what kind of predator is approaching. The semantic content of vervet monkey alarm calls was shown in important work by Robert Seyfarth and Dorothy Cheney (1980). Their experiments revealed that the vervet monkeys classify and "read" the message purely on the basis of its acoustic structure, even when they are deprived of any visual clues. They found that vervet monkeys achieve major changes in signal function by changing frequency peak: calls with an early peak serve an integrative or cohesive function for the group; calls with a late peak indicate a state of arousal (sexual or agonistic).

Larger forest-dwelling primates have succeeded in exploiting vocalization over long distances, despite the severe constraints imposed by a forest environment. Mangabeys in the Kibale Forest in western Uganda can be heard (by the human listener) from a distance of 500–600 meters through dense forests and, at certain times, even from as far away as 1,200 meters. The loud calls of chimpanzees, known as "pant-hoots," are also audible for more than 900 meters, as Peter and Mary Waser (1977) have reported. The "long call" made by orangutan males, a spine-chilling roar, is audible for at least 1 kilometer. These calls may advertise presence but may also convey specific information on local dialects, as Andrew Marshal, Richard Wrangham, and Adam Arcadi have recently found. Although they studied captive chimpanzee populations it was clear that the pant-hoots varied from one group to another and that learning occurred in order to communicate (Marshal, Wrangham, and Arcadi, 1999).

Gibbons (*Hylobates* spp.), also forest dwellers, are among the most conspicuous vocalizers of all primates. They are territorial and monogamous. Not unlike some bird species, they sing duets in which the female usually takes the lead. These are extensive vocalizations and the skill involved in producing them lies mainly in the coordination of the "song" in the duet. The song of gibbons also tells of the length of the pair relationship—inexperienced pairs usually have problems with their duetting coordination and may not finish their song. The sounds of these duets can be heard for miles across the rainforests of southeast Asia. In the case of gibbons, the song is less concerned with maintaining contact with individuals than with protecting territory—advertising their presence in a patch of forest.

Nonvocal Sounds

Sound signals are not confined to vocalizations. Mammals may be better equipped to make sounds with their limbs than birds. And there is plenty of evidence that the limbs are used extensively in a large variety of contexts, such as territorial defense, courtship, and identity marking. For instance, banner-tailed kangaroo rats *(Dipodomys spectabilis)* use individually distinct foot-drumming signatures to communicate their identity to territorial neighbors. They can also discriminate between the foot-drumming signatures of neighbors and strangers. Jan Randall (1994) thought that familiarity among neighbors promotes a stable social organization in this solitary, nocturnal rodent. Foot drumming is also used in the threat displays of nocturnal lower primates (prosimians). Gibbons shake branches, break them off, and drop them in what has been called a brachiation display. It is a stunning and very noisy affair, achieved solely by using branches. Alpha male chimpanzees (the dominant males) like to make noise too when they display, breaking branches and bashing them as they run. Presumably the noise reinforces perception of their physical strength. Gorillas use chest beating, first documented in detail by George Schaller in 1963. Lip smacking is another nonvocal sound that plays an important role in some apes and monkey species during allogrooming (grooming other individuals). It is used by an approaching monkey to indicate its peaceful intention and then maintained during the process of allogrooming.

Early research on primates attempted to catalogue primate vocalizations, particularly those of the great apes (chimpanzees, gorillas, and orangutans). The vocalizations of orangutans have not been studied systematically, but studies on chimpanzees and gorillas have had some interesting results. When Peter Marler and Richard Tenaza compared the vocalizations of gorillas and chimpanzees in 1977, they found that "the most striking conclusion to be drawn from the data is the surprising degree of correspondence between the two species in the rank order of use of corresponding calls." By this they mean that there were important similarities between the species in the order of most-used to least-used vocalizations. This is an important point because the social organization of chimpanzees is not at all like that of gorillas. These similarities may indicate the animals' common evolutionary origins, cultural transmission by learning, or the fact that both species are subject to similar environmental constraints on vocal transmission. Almost certainly all three factors exert an influence.

SCENT DEPOSITS AND OLFACTORY MARKERS

Chemical communication is a widespread form of communication among mammals. It has been recorded in rodents (mice, hamsters, rats, voles), marsupials (koalas, sugar gliders, opossums), ungulates (horses, deer), dogs, primates (Old and New World monkeys), and even elephants. Chemical communication is much older than mammalian existence. Fishes, amphibians, and reptiles also use chemical communication, in alarm signals and in courtship, in kin recognition and territorial defense. Members of the most ancient marsupial family, the Didelphidae (such as the gray opossum), show extensive scent-marking behavior. They also show estrous synchrony, caused by odors known as pheromones, and estrous activation, also triggered by pheromones (Fadem,

1985). A pheromone is a chemical, or mixture of chemicals, released into the environment by one animal that causes a specific behavioral or physiological response in another animal. As well as the species named above, many insects use pheromones. Olfactory communication has a long evolutionary history.

The scent-releasing glands of many species are larger in the male than in the female, and in both sexes are more highly developed during the breeding season. Species with well-developed scent glands tend to be polygynous, and territorial males tend to scent-mark more than non-territorial males. We have seen that communication tends to become more complex with the complexity of the group, and territoriality is another variable that adds to the range of communicative needs. Territoriality needs to be communicated, and hence scent marking becomes a constant activity in the effort to maintain a territory.

Chemical signals seem to occur most frequently among species that are subject to predation and have a limited home range. In primates, at least, it is known that the most developed system of chemical signaling and communication is found in those species most subject to predation, especially those that are nocturnal and/or arboreal. Such arboreal species include New World monkeys, such as marmosets and tamarins, as well as Old World nocturnal species, such as prosimians. All these species rub their scent glands on objects in their environment, marking them with their scent. Prosimians also mark themselves with urine: they urinate onto their hands and rub the urine into the fur.

Primate species that are largely terrestrial (living on the ground) and diurnal (active during the day) tend to rely on chemical signals to a lesser degree. For instance, apes do not have well-developed olfactory lobes (located in the brain) and do not rely on scent marking. Apes are large and/or live in strong groups and generally have few predators. They are mobile and may forage over large areas. All these characteristics have led to the belief that intense reliance on olfactory signals occurred in ancient primate species and that use of olfactory signals has been largely superseded over time to include other senses considered to be more suitable for communication. However, apes do rely on olfaction to signal reproductive conditions and olfaction may play a greater role in other aspects of their social behavior than we currently realize.

The line between intentional and unintentional communication by odors may be fluid. For instance, individual A (female) of a species may give off scents that will entice individual B (male) to mate with her. Individual A may not have produced the olfactory signal intentionally, it being merely a result of her changed hormonal state. Individual B has learned to interpret the signal correctly. But let us say that individual B has the choice of several females that are emitting signals similar to those of individual A. Individual B can therefore choose according to the quality and type of smell. Whether the choice is made intentionally or unintentionally is likely to vary with the species. Males of many species, avian and mammalian alike, have developed very elaborate strategies to ensure that they will succeed in getting a mate. Olfactory, visual, and vocal communication is used to achieve this end. Moreover, mate choice is not confined to males. In the great majority of species, females do the choosing and the male the displaying and competing (see McFarland, 1985).

Scent may be used to make territorial claims and to defend the territory, and odor signals require more than merely releasing a specific odor from the animal's body. There is a good deal of work involved in scent-marking a territory. Gisela Epple found in the 1970s that olfactory signals have a complex set of communicative functions in the life of the common marmoset. Marmosets have several glands that release different odors, under the chin, on the chest and in the anogenital region. First, there is intragroup communication, including sexual communication, regulating social relationships among adults and infant-adult relationships. Second, olfactory signals are important in intergroup communication concerned with territorial defense and the formation of new groups. Third, olfactory signals help to maintain orientation in the environment. Further studies by Epple (1988) and her colleagues on saddleback tamarins *(Saguinus fuscicollis)* and cotton-top tamarins *(Saguinus oedipus oedipus)* have shown that their olfactory communication may fulfill a range of communicatory functions similar to the range seen in the marmoset.

While olfactory signals in the urine play an important but balanced role in some monkey species, other mammals depend on olfactory cues almost exclusively. For instance, interactions between house mice depend to a significant extent on olfactory communication. An experiment conducted by Jane Hurst (1993) and her colleagues showed that male house mice remain tolerant toward subordinate males largely because the subordinates leave urine deposits in spots and streaks across the entire territory (substrate odor deposits). Resident male mice, both dominant and subordinate, behave aggressively toward subordinate males that do not deposit fresh odors in the group's home territory. It is obviously important to be known in an olfactory capacity to dominant males. The quality of urine in subordinate males is different from that of dominant males. Hurst and her colleagues suggest that regular urine markings by subordinate males is an efficient system, allowing territorial males to concentrate their defense on intruders. It appears that mice and quite a number of other mammalian species need to supplement their visual displays with cues from other sensory modalities, in this case smell. The substrate marking by a subordinate male reassures the dominant male that no attack is planned on his status and territory.

In the European rabbit (*Oryctolagus* spp.), chin marking is one of the most conspicuous forms of olfactory communication. Robyn Hudson and Thomas Vodermayer (1992) found that secretions from the chin gland were used by females as a sexual advertisement but also served nonsexual functions. Female rabbits are able to discriminate between chin marks from different animals on the basis of the donor's hormonal state. The researchers conclude that chin marking may also play a role in the establishment and maintenance of group identity. Group stability and territorial stability may thus be served by extensive use of olfactory signals. Michael Stoddart (1992) has found a similar use of odors in the social behavior of marsupial sugar gliders *(Petaurus briceps)*.

We have not yet raised the possibility that signals may get lost, misread, or overlooked. The issues of selective attention and selective memory may be of great importance in communication and constitute a study in their own right. Suffice it to say here that a recent study on golden hamsters (*Mesocricetus auratus*) by Robert Johnston (1995) and his colleagues has drawn our attention to the fact that not all messages have a recipient and that, for some species, this seems to have evolved by design. The researchers tested hamsters for their responses to the partially overlapping scents of two individuals to see whether the hamsters would be able to identify both individuals. They found that the hamsters remembered only the scent mark on top—the one deposited most recently—even if the other scent had been identified before in a separate test. If scents are "read" only selectively, we would have to adjust our thinking about communication to include as part of the process selective detection and selective perception. We would have to realize that individuals may be able to focus their attention on a particular odor relevant at a particular time and in a particular context, neglecting other odors that are present.

TACTILE SIGNALS

A good deal of communication can also happen by touch. Grooming in mammals is an important gesture of intimacy and closeness. It reinforces pair bonding, as it does in birds, and in certain primate groups, such as rhesus monkeys and baboons, grooming is associated with status within the group. Dominant members of the group are groomed by subordinate ones. Sometimes, there are lines of animals each grooming the next one in the row. Tactile communication in baboons and bonobos is often used to signal appeasement, reassurance, and loyalty. An animal's intention to groom usually has to be advertised so that the individual being approached is assured of the peaceful purpose. In baboon groups, the approaching individual smacks its lips loudly and then continues the lip smacking throughout the grooming process.

Another form of body contact is embracing. Obviously this form of tactile contact relies on the existence of limbs that can do the embracing. We find this form of communication largely in monkeys and apes, although it does occur in mating frogs and toads. Hugging, cuddling, and cradling are activities not confined to mother-infant interactions; they are found among nonrelated animals, even of those of adult age. Bonobos may be unusual, even among apes, in that they use "loving" tactile contact (all sorts of tactile activities, including sexual ones) for settling conflicts within the group. Notably, as Frans de Waal and Frans Lanting (1997) point out, it is mostly the females who maintain peace by means of physical contact with each other. Lip touching of two conspecifics—kissing—may be simply a friendly greeting or it may be an overture to sexual advances.

Orangutans, too, use touching extensively in certain social contents. Mother-infant relationships, which are very intense and long-lasting, are established by close physical contact (as we describe in detail in our book *The Orang-utans,* 1999). Juveniles and even adults (usually females) continue to use touch as a form of communication, often without eye contact. Juveniles often walk along holding hands. Although orangutans are largely solitary, rather than living in groups like the other apes, the sense of touch in personal relationships continues to play a role through adult life. It obviously features in sexual behaviour.

Elephants use their trunks extensively to communicate with each other. They have very sensitive skin and the trunk needs only to glide gently over the body, or touch the trunk of another, for a message to be conveyed. The trunk is used to help baby elephants stand up and walk when the herd is moving to new feeding grounds, and also for reassurance and many other subtle communications.

Dolphins and whales use touch as a form of communication for many of the same reasons (as far as we know) that touching is used in other species. They nuzzle each other with the snout or swim alongside each other, brushing along the skin.

Many mammalian species use licking as a form of reassurance, as an expression of bonding, or as a signal of status. Dogs and related canids, for instance, use licking extensively and in a variety of social contexts, as Michael Fox (1971) showed in his book on canid behavior. Dogs go through extensive daily rituals of reassuring each other and of reconfirming the status of the lead bitch. They may do this by touching the other dog's nose or licking the other's snout. Conflict resolution is usually swift, whether it is fierce or friendly. In extreme cases, a dog may be expelled from the pack or even killed (depending on the species), but dogs usually resolve conflicts in a conciliatory manner. In conflict resolution, licking is directed behind the ear, on the neck, and, if allowed, in the anogenital region. Small bites, shoves, and pushes are all part of a gentle and friendly communication.

RECOGNITION OF INDIVIDUALS

Can animals recognize conspecifics as individuals? Do they relate to individuals in a specific way? Or do they just respond to key markers of categories, such as plumage color, a particular scent, or a specific vocalization, that are sufficient to trigger "familiarity" or "stranger" status? And are such questions meaningfully applied to all animals or only to some? Those who hold a mechanistic view of animals would certainly regard mere reaction to specific markers as sufficient for survival and would say that animals' abilities stop there. But many people think at least some animals have more extensive abilities. Stanley Cohen (1994), for instance, draws attention to the intelligence and capabilities of pet dogs (see also Fox, 1971). Most pet owners are convinced that their pets can identify them as individuals. This recognition of the owner is part of the close relationship that is formed between owners and pets (see Chapter 8). But to what extent does this recognition of individuals apply in the wild, and how exactly do animals recognize each other? It is not always easy to determine scientifically what signals animals might use to achieve recognition of individuals.

For animals to recognize individual conspecifics as unique entities, most researchers assume, they need to have a memory of each individual, a representation composed of a variety of key markers, or "integrated, multi-factor representations." This assumption was tested in golden hamsters *(Mesocricetus auratus)* by Robert Johnston and Paula Jernigan, who showed that golden hamsters respond to individually distinctive signals on the basis of the meaning (or the referent) of the signal. In their experiments, male golden hamsters were exposed repeatedly to the scents of females in estrus, and the males clearly could distinguish between a familiar female (an individual) and a strange one, and also could distinguish between two odors of the same female while still attributing them to the same individual. Johnston and Jernigan suggest that this result indicates the importance of higher-order, cognitive processing in the social behavior and communication of hamsters because the animals categorized stimuli according to their significance and not strictly by their sensory characteristics (Johnston and Jernigan, 1994).

Recognition of the alarm calls of different conspecifics also seems to be important factor in the recognition of individuals, because some individuals signal the presence of predators more reliably than others. Unreliable signalers that "cry wolf" too often can be ignored if they are recognized. Some recent research on the ground squirrel *(Spermophilus richardsonii)* has shown that this may be the case. James Hare (1998) recorded the alarm calls of different squirrels and then played them back to selected individuals in their natural environment. He found that a squirrel no longer attended to hearing the same individual's alarm call after it had been played back four times. Habituation had occurred. Then he played back either another alarm call by the same individual or the alarm call of another individual. The squirrel became more vigilant after hearing the call of the new individual but not after hearing another call by the first individual. It was able to distinguish one individual's call from another's.

The ability to make distinctions between individuals would also help group-living animals observe rules in established social hierarchies and in other social relationships. For example, individual recognition in rhesus monkeys has been shown to be very sophisticated. Vocalizations by a dominant member of a group may require a different response than vocalizations by a subordinate. The maintenance of group structure and (in the case of alarm calls) even survival may depend on the ability to distinguish the vocalizations of dominants and subordinates. Habituation to alarm calls by trusted/senior individuals of a group could threaten survival.

Not only do receivers of calls distinguish individuals but they also respond to the calls according to the caller's relationship to themselves. In a social system in which the mother's relatives (the matrilineal line) play a significant role, categorization of other animals' calls by lineage might be important. Playback experiments using calls of unrelated and related individuals have been conducted by Drew Rendall and his colleagues (1996). They have shown that female rhesus monkeys *(Macaca mulatta)* respond significantly faster and longer to contact calls of matrilineal relatives than to calls from other relatives and from nonrelatives. Their study demonstrates that rhesus monkeys are able to distinguish unrelated individuals from kin. But even after such experiments have been conducted under controlled conditions, it is not certain whether true recognition of individuals has occurred because other cues (such as the location of the individual) may assist in identification. The researchers point out, however, that the capacity to recognize vocalizations of individuals and kin represents an important adaptation in long-living primates, who have complex social relationships between individuals.

In large social groups, individual conspecifics may need to be known to each other. The question is whether individuals in a group would at once recognize an intruder and whether they could do so by visual information alone. According to recent studies (for example, Parr and de Wall, 1999), chimpanzees can perceive similarities in the faces of conspecifics who are related to individuals they know but are unfamiliar to them. They can recognize kin by facial features (Parr and de Waal, 1999). This ability was discovered when researchers showed pairs of photographs of conspecifics to the chimpanzees. The chimpanzees were able to recognize relationships between mothers and sons but not relationships between mothers and daughters, a fact that is likely to have social implications for chimpanzee society.

As we have seen before in the discussion by Janik, Dehnhardt, and Todt (1994), dolphins use signature whistles that identify individuals. There seems little doubt that mammals recognize each other individually, but determining exactly how they do so in each species requires much more research.

HUMAN COMMUNICATION WITH NONHUMAN PRIMATES

The effort to communicate effectively with nonhuman primates via language or a system of symbols has generated much innovative research. The phylogenetic affinity of the great apes to humans seems to make it possible to devise ways of bridging the gap between animals and humans. If real communication with the great apes could be achieved, we would gain much information about their personalities: their thoughts, memories, wishes, fears, and a host of other things that cannot be deduced from observation alone or that are not unambiguously measurable. Some notable researchers have tried to create that bridge by including apes in their personal lives, raising chimpanzees and gorillas as if they were their own children. Others have moved into the natural environment of the apes, staying in close proximity to them until they were finally tolerated or even accepted by the group. These pioneering research efforts led to a sense that some real communication had taken place, based on trust and mutual respect. Many new insights were gained in the process, and if we can speak today of awareness and consciousness in animals, we can do so largely because of the research on communication undertaken with great apes.

However, researchers have gone down a number of blind alleys. For instance, many studies of vocal abilities in primates were driven by the wish to understand the origin of human language rather than the workings of animal communication. Roger Lewin (1991), for example, argued that chimpanzees may hold the only key to the origin of human language. These studies assume the superiority of vocal communication. But vocal communication may not be a superior form of communication; it may simply be the one we understand best and one that has served the evolution of human primates extremely well. Also, it is often assumed that species close to humans—primates—should show more evidence of vocal learning (higher plasticity in their development) than species that are more distant from us in evolutionary terms. This is not the case. The development of vocalization and vocal learning have been shown to exist in songbirds, as we noted earlier and as we will discuss further in Chapter 6. In fact, overall, less is known about vocal development in primates than in birds, even though we now have some very detailed knowledge of the vocal communication systems of the great apes. Although, as we have shown, there have been very successful attempts to teach birds to speak, attempts to teach apes to speak have failed, because the vocal apparatus of apes is not constructed to produce human speech sounds. Apes can communicate with humans by using sign language or symbols.

ETHICAL QUESTIONS IN COMMUNICATION RESEARCH

Investigations of animal learning, adaptation, and communication have often ignored the animal as a whole organism, especially when only one aspect of the animal's behavior was being studied. Meredith West and her colleagues (1997) have recently raised ethical questions about experiments on vocal learning in birds and primates. They argue that in the past researchers, in their desire to establish the parameters of learning, often used methods of testing that would now be considered unacceptable. In some early experiments, for instance, monkeys were kept in very small chambers for an entire year with no physical access to other animals. Similar ethical issues have arisen in studies of birdsong, where some birds are often kept in prolonged isolation in order to control the experiment.

Although conditions for animals have improved greatly over the past 20 years, living conditions for experimental animals inevitably involve deprivations. Most experimental species are kept in sterile environments and confined to cages where they have little to do. These problems, ethical and experimental, have of course been recognized, and many studies have attempted to remedy the situation by improving the physical environment of captive animals or by complementing laboratory studies with field studies. The problem is that there is no perfect system of studying animal communication that is completely noninvasive, involves no deprivation, and is scientifically unassailable. In the natural environment, controls are more difficult to establish and hence results may be more unreliable. The laboratory setting, by contrast, allows the establishment of controls, but may distort results by its very artificiality. West and her colleagues found that differences in social and physical settings in cage and aviary tests could lead to the display of different levels of competence (or the lack thereof) in social and communicative skills.

CONCLUSION

The topic of learning and communicative competence invites further comment, and we will explore it further in the next chapter. Suffice it to say here that there appear to be many aspects of communication with a long evolutionary history which we share with birds and mammals alike, whether these be body postures and displays, facial expressions, or vocalizations expressing alarm, reassurance, and anger. It is these that researchers have tended to recognize most readily and that are being catalogued. The challenge is to recognize and study the complexities of species-specific forms of communication which we humans do not share or do not share fully.

Chapter Six

LEARNING TO COMMUNICATE

In earlier chapters we discussed some of the varied patterns of communication used by different species. In most cases we focused on the communication patterns of adults; very often these patterns are not present in the behavioral repertoire of infants or juveniles but develop as the animals grow up. This development is partly due to maturation as the animal gets older, a process dependent on the unfolding of its genetic program (which is read out from the genes passed on through generations), and partly due to experience and learning. These two processes are often regarded as separate, but they are not. At every stage of development, maturation, experience, and learning interact.

Let us consider a familiar example that is very relevant to communication. The maturation of the reproductive organs and the consequent release of sex hormones has a major impact on vocal communication in many species because the hormones affect the growth of certain parts of the brain and the vocal apparatus—the larynx in mammals and the syrinx in birds. In the human male, changes in the larynx cause the voice to deepen. In birds, syrinx growth often coincides with the emergence of new vocalizations. For example, roosters start to crow when they approach sexual maturity because that is the time when their sex hormone levels rise. If the sex hormone, testosterone, is injected into young chicks, they will crow, but they will sound like very squeaky roosters because the syrinx has not yet developed enough to make a full crowing sound.

In songbirds, it is known that the sex hormones affect the development of certain structures in the brain that are used to control singing. As Fernando Nottebohm (1989) has shown in his research on the canary *(Serinus canarius)*, certain regions of the forebrain enlarge as the amount of testosterone circulating in the blood increases. The genetic program for development plays a part in determining this sexual maturation process but experience also contributes.

To continue with the songbird as an example, the season of the year provides the trigger for the development of the sexual glands. As spring approaches days grow longer, and this increase in daylight is the stimulus that causes enlargement of the sex glands and increased release of sex hormones into the bloodstream. In males, these changes in turn cause the regions in the brain that are used for singing to enlarge. Once this has occurred, the bird is able to sing the songs special to the breeding season.

In establishing the canaries' songs, learning is also involved. Male canaries elaborate on their songs each year; they learn from hearing themselves and other canaries, and they remember their own songs from year to year. Thus the song that each bird produces has been determined by the interactive effects of its genetic program, the experience of increased day length, the level of the sex hormone testosterone, and learning. The genes and the environment interact to determine the song of each canary, and as might be expected, each individual sings a different song. We discuss this interaction in more detail in Chapter 7. Here we note that there are differences between avian species in when birds sing and whether only males sing. As we mentioned earlier, both male and female Australian magpies sing and they do so all year round. So far, there has been little research on species that sing all year round and in which both sexes sing.

Not only must a bird know how to sing, but it must also know in exactly what place and at what time of day it is advisable to sing (singing in another bird's territory would provoke attack). The bird must also know which individual to direct its singing toward. Similar criteria apply to all animal species and also to other forms of communication. Since communication is social behavior, it is not surprising that there are many different aspects of communication that have to be learned. First, we discuss learning to produce vocalizations and then we consider the importance of learning when, where, and how often to communicate.

SONG DEVELOPMENT IN BIRDS

There has been much interest in the study of song development in birds. Three kinds of evidence indicate that a vocalization is learned. The first kind of evidence is the development of abnormal vocalizations in birds that are raised in isolation from conspecifics, so that they never hear the vocalizations of their own species. The second is the abnormal development of vocalizations in individual birds that have been rendered deaf early in life. This is not a procedure that would be approved for experiments conducted nowadays, but it was used three decades ago and we report the results because they contain valuable information that should not be lost. The third kind of evidence is that of vocal imitation or mimicry of the vocalizations of other species and of sounds in the environment.

There is no evidence that vocal learning occurs in the Galliformes (chickens, turkeys, quails, pheasants) or the Columbiformes (pigeons), but it does occur in the Passeriformes (in the large number of different species of songbirds known as oscines, but not in all the Passerines), Apodiformes (hummingbirds), and Psittaciformes (parrots). There are, of course, numerous species in each of these categories. The ability of parrots to imitate human speech, and sounds such as the creaking of doors and the noise of a bottle being opened, is well known (see our discussion of the parrot Alex in Chapters 2 and 3 and also see a 1975 paper by Dietmar Todt). Many songbirds also mimic sounds in their environment and, when hand-reared, will mimic human speech. We discussed this special kind of learning in detail earlier; here, we are more concerned with the first and second kinds of evidence showing that avian vocalizations are learnt.

Learning of vocalizations is characteristic of those oscines that have complex songs as well as those with local dialects (variations in their vocalizations from one region to another). One of the latter species is the chaffinch *(Fringilla coelebs).* Some time ago, William Thorpe conducted some very important experiments in which he hand-reared male chaffinches in isolation from other members of their species and then studied their song development. In this species only the males sing. When the chaffinches became adults, the songs of the hand-reared males were very different from the songs of wild, adult chaffinches, although they were of roughly the same length, covered roughly the same range of frequencies (pitches), and were subdivided into packets of sound in somewhat the same way. It was as if the males reared in isolation retained a template for the song but, lacking social experience with their own kind, were unable to learn the species-specific song (Thorpe, 1961).

As confirmation of the importance of learning, it was found that, if the male chaffinches are played a tape recording of a chaffinch song as they grow up, they learn that song and produce a song that is almost identical. The same has been shown in other species, such as the song sparrow *(Melospiza melodia)*. Other species show a similar dependence on hearing another bird singing early in life but, unlike the chaffinch, they need to interact with a living bird—simply hearing a tape recording of the sound is not sufficient. The Australian zebra finch *(Taeniopygia guttata)* is an example, as we discuss further below.

Even establishment of the template of the song requires some learning, but in this case the bird learns by hearing itself. This was first demonstrated in the song sparrow by rendering birds deaf early in life. The deafened birds developed songs that were entirely different from the songs of adults in the wild. They exhibited either no evidence of a template of the species-specific song or a very crude template, much less structured than the template that develops in isolated, hand-reared song sparrows. Their songs were even more abnormal than those of hand-reared members of their species. The same result was found in deafened chaffinches and other songbirds.

Birds learn to sing very early in life. Certain vocalizations are learned more readily than others: each species selects particular vocal patterns to memorize. Genes seem to determine this initial selection of the first types of song to be memorized (see Marler, 1991 and 1997, for more detail). Later learning shapes further selection of songs to memorize.

There are several distinct phases in the development of song, which we will illustrate by discussing the chaffinch. Soon after hatching, young chaffinches produce begging calls to which the parents respond by supplying them with food. By the time of fledging (at about the age of 5 weeks), these calls have been replaced by rambling, soft vocalizations, referred to as "subsong." Subsong is often produced when the birds are dozing or perching quietly. The bird runs through a whole series of different notes and the sequence can be very long. Subsong occurs in many species and, as Peter Marler (1970) has pointed out, it is remarkably similar to the babbling sounds that human infants make when they are acquiring speech. Both the subsong of birds and the babbling of humans provide auditory feedback—the individual hears the sounds that it is making—and so self-learning is probably occurring. The equivalent of subsong also occurs in parrots during a stage of life when they are practicing their learned vocalizations. Irene Pepperberg (1991) has reported on what she calls "solitary sound play" by the parrot Alex when he was being taught new vocalizations. At these times, Alex produced sounds that were similar to, but not exactly the same as, the new words that he was learning.

The subsong of the chaffinches comes to imitate parts of the parents' song, although not precisely. This imitative song is called plastic song because it is still variable and has not yet developed into adult song. The bird's song practice subsides during the winter. The next spring singing begins again, and this time it is subsong interspersed with plastic song. One month later, the song crystallizes into "full song." The same pattern occurs in many other species of songbird, including song sparrows, cardinals, and buntings. The Australian magpie does not follow this linear pattern of song development to crystallized song, although it does have a plastic song.

As mentioned above, chaffinches exposed to an adult's song during their early development learn that particular song and produce a copy of it when they themselves become adults. A short exposure to the adult's song during the first few weeks of life is sufficient; the bird will then reproduce that song in adulthood even without further exposure to it for many months. This shows that there is a sensitive period in the chaffinch's early life during which song learning occurs. Chaffinches, and many other songbirds including zebra finches, learn song early in life, and when they become adults, they do not change their song. Canaries also learn their songs early in life, but they are able to change their songs from season to season when they are adults. It appears that they go on learning throughout their lives.

Learning in adulthood is not limited to canaries: some parrots have been reported to learn new sounds when very old. We have seen a parrot (a galah) more than 60 years old learning new words after moving into our household. These words were the names of two of our dogs. Not only did the galah learn to imitate the words but he uses them only when the dogs are missing or expressing aggression to each other—he has never used the words out of context and he does not use them very often. This ability to change vocalizations in later life also appears to apply to the natural vocalizations of parrots.

The same ability has been shown in budgerigars by Susan Farabaugh and her colleagues. These researchers found that individual caged budgerigars *(Melopsittacus undulatus)* changed their contact calls so that they resembled more closely those of another budgerigar caged alongside. The budgerigars showed mutual learning of each other's calls. By imitating each other, they converged their calls so that they became more alike (Farabaugh, Linzenbold, and Dooling, 1994).

The ability to continue to learn in adult life does not, however, lessen the importance of the sensitive period for vocal learning in early life. There appears to be a window that opens in early life and allows the bird to learn a wider variety of songs than it can learn either before or after that sensitive period. This was demonstrated clearly by experiments conducted by Donald Kroodsma in the late 1970s (Kroodsma, 1978). He exposed long-billed marsh wrens to a large number of different songs, a few songs every 3 days. The period of exposure began soon after hatching and continued until about 85 days of age. The wrens learned very few of the songs they heard before about 25 days of age, although the number they learned increased from 10 days of age on. The best period for acquiring a variety of songs was 25–55 days after hatching, but there was a period of less learning around 40–45 days; from 55 to 80 days there was a decline in the number of songs learned, although the exposure to different songs was just as various throughout this entire period of time. The results show clearly that there is a sensitive period during which the marsh wren learns new songs (see Figure 6.1).

The ending of the sensitive period may depend on changing hormone levels; an injection of testosterone into zebra finches before the normal end of the sensitive period has been shown to curtail the learning of new songs. This effect contrasts with the onset of song production in the next spring season, when testosterone levels rise and the zebra finch sings the songs it learned earlier during the sensitive period. It would appear that high levels of the hormone testosterone crystallize the song so that no new learning will occur. The same hormonal condition also stimulates the singing of the songs that have already been learned. In adults the seasonally fluctuating levels of testosterone also affect song: a recent study by Troy Smith, John Wingfield, and their colleagues has found that male song sparrows sing songs that are more variable in the autumn, when their testosterone levels are low, than in spring, when the levels are high, although the same repertoire of songs is sung in both seasons (Smith et al., 1997).

Sensitive periods for learning are not restricted to song learning or to avian species. There are sensitive periods for learning other behaviors, such as the sensitive period for forming social attachments by the process of imprinting. There may also be a sensitive period for learning language in humans, as isolated examples indicate but do not prove. There is the documented case of Genie, a human child who was denied any form of normal social or linguistic experience for the first 13 years of her life (re-

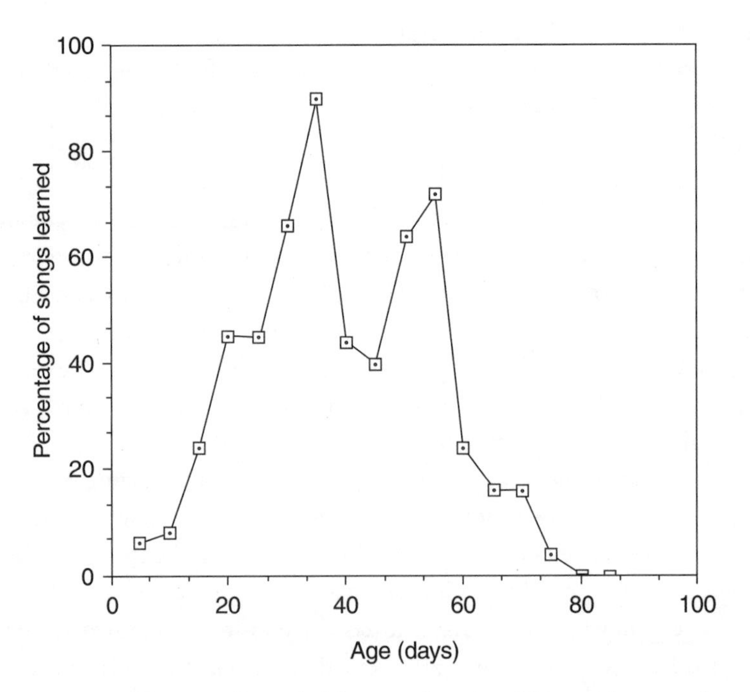

FIGURE 6.1 The sensitive period for song learning. This graph shows the percentage of songs learned by long-billed marsh wrens *(Cistothorus palustris)* at different ages after hatching. The birds were exposed to a large number of song types during this period but not all at once. Over each sequence of 3 days they were exposed to only a few of the songs.They copied more of the songs heard between days 25 and 55 than at other times, although fewer songs were learned between days 40 and 45. (Calculated from the data of D. Kroodsma in Slater and Jones, 1997).

ported by Curtiss et al., 1974). Genie never acquired full speech capacity, and the abnormality of her first 13 years, and indeed the years that followed, may have contributed to this failure of recovery. There have been similar cases with comparable results, but all these cases of isolated humans have had highly abnormal aspects that confound the conclusions that can be drawn from them. This might also be said of the birds that were reared in isolation. It is not possible simply to take something away—in this case, the bird's normal experience of hearing song and experiencing other social interactions—without causing unexpected effects on behavior in general. Development is not a simple process from which one can extract a single aspect without causing unpredictable effects.

The inability to learn song or language after an early life of social deprivation might be an aberrant outcome not directly related to the simple subtraction of a normal aspect of social experience. There is a well-known example that may help to illustrate this, and that is the experiments performed by John Paul Scott (see Scott and Fuller, 1965), in which he raised dogs in isolation from the time of their birth. When they were brought into contact with people later in life, they behaved in very abnormal ways, one of which was to rush and bite at the flame of a cigarette lighter. Being raised in isolation led not just to an absence of some patterns of behavior but to the emergence of behaviors never seen before. These results alert us to be cautious in interpreting experiments in which animals are raised in social isolation, as in the studies of the songbirds.

SINGING TUTORS

As we have said, chaffinches and marsh wrens will learn songs from tape recordings played to them during the sensitive period. But zebra finches and some other species cannot learn from tapes. They need to see and interact with another bird of their own species at the same time that they hear it singing. Even if the other bird sings within their earshot but is hidden from their view behind a screen, they will not learn their species-specific song. It is some aspect of interaction with the singing bird that counts, as was shown by Patrice Adret (1993). He was able to train zebra finches to sing by allowing each bird, caged alone, to peck at a key to turn on a tape recording of a zebra finch's song. By turning on the tape recorder the birds were able to interact with the artificial "tutor." The birds trained in this way pecked at the key many times in a day in order to hear a small segment of song (only 15 seconds in duration), and they often flew up and down in front of the loudspeaker as the tape recording was playing. They learned to sing the same song as the tutor and produced it when they became adults. Control birds exposed to the same tape recording of song but in a passive way (they could not turn it on themselves) did not copy the recorded song. This research shows that some form of interaction with the tutor is essential for learning to occur, no matter how unusual that interaction is.

Peter Slater and his colleagues have shown that a zebra finch may prefer to learn the song of its own father, but this is not at all a straightforward process (Mann and Slater, 1994). Male zebra finches usually learn their songs in the second month of life. In the experiments conducted by Slater, the young zebra finches were housed with their parents until they were 35 days old, and so each could hear its father's song over this period, which precedes the sensitive period for song learning. Then each young bird was caged separately in the central part of a cage with three partitions (Figure 6.2). From day 35 to day 100 of life each bird was exposed to singing birds placed in the compartments on either side. In the first experiment, an adult male was housed alone on one side and an established pair of birds on the other side. Neither male was the parent of the young bird in the central cage. The young birds learned to copy the song of either the single or the paired male, but they preferred to learn the paired male's song rather than the single male's song.

In another experiment, Slater and his colleagues exposed a young zebra finch to his father caged alone on one side and his mother caged with an unfamiliar and unrelated male on the other side. Of the 13 birds tested in this way, 10 copied the song of the unfamiliar male housed with the mother, 2 learned their father's song, and 1 learned equally from both tutors. Thus the preferred tutor is the male paired with the bird's mother, not the actual father, even though the father's song had been heard for the first 35 days of the young bird's life.

The final experiment gave the young zebra finch a choice of learning from his father housed with an unfamiliar female on one side or from an unfamiliar male housed with his mother on the other side. Of the 16 birds tested, 10 learned to copy their fathers and 6 copied the unrelated

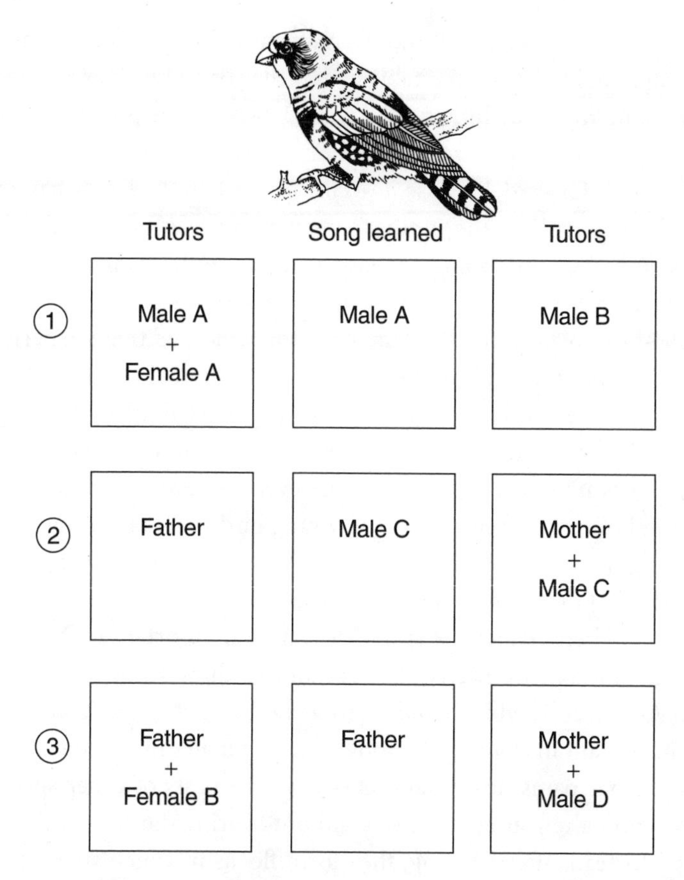

FIGURE 6.2 Song tutors. This figure summarizes the results of experiments on zebra finches by Nigel Mann and Peter Slater (1994). Each young male zebra finch was placed in a central cage with potential song tutors on either side. The song copied by the learner is indicated in the central box. Males A to D and females A and B are unrelated and unknown to the young male who is learning his song. Note that the young bird prefers to learn the song of a male paired with a female and also to learn the song of his father if the father is paired. But the young bird prefers to learn the song of an unfamiliar, unrelated paired male rather than his father's if the father is not paired.

tutor. In this case the birds showed a preference for the father's song. This preference could have been established by the young bird's exposure to its father's song from hatching until day 35. However, the preference for the father's song after exposure during the sensitive period for song learning is not straightforward, because it occurs only when the father has a partner. If he is unpaired, the mother's partner is preferred over the father. Hence both the mother and the father influence the young bird's selection of a tutor.

Preference for a particular singing tutor is not, as these experiments show, a simple matter. Another experiment found that tutors that are more aggressive and interact with the young zebra finch by pecking and chasing him are copied more than less aggressive ones (Weary and Krebs, 1987). We assume this happens because even aggression increases the interaction between the tutor and the young bird that is learning, but it is surprising that a somewhat punitive interaction would be effective. This is an area that deserves further investigation.

The bond between female and male is not unimportant in the learning of song. Research by Meredith West and Andrew King (1988) on the North American cowbird *(Molothrus ater)* shows that the female can be most important in shaping the song of young males in this species. Since this species is parasitic—it lays its eggs in the nests of other species, as cuckoos do—the young are raised without hearing the calls of their own species. Instead, after fledging, they form flocks in which singing by the males is shaped by the young females. The female performs a display of "wing stroking" when she is attracted by the song of a young male. This apparently reinforces the male because he is more likely to sing the same song again if the female has performed this display. Therefore, in time, the songs of the males become matched to the preferences of the females. The females train the males to sing the correct song even though they do not sing themselves. If the males are put into a flock with females of a different group of cowbirds, they learn, also in response to wing stroking by females, to sing the song of that group instead of their own.

THE CULTURAL TRANSMISSION OF SONG

The transmission of song from one generation of an avian species to the next by the process of learning has been viewed by ethologists as cultural transmission. In fact, several researchers in the field (e.g., Slater and Ince, 1986; Trainer, 1989) refer to the changes in song that result over time as the song is passed from generation to generation as "cultural evolution," as distinct from genetic evolution. It is thought, however, that the changes that occur as song is passed on are due to errors in copying (the bird does not produce an exact copy of the song that he heard) rather than being innovations on the part of the singer or some new form of adaptive behavior. Inevitably, small errors will creep into the copied song over time, but nevertheless copying is surprisingly accurate. In addition to this source of change in the song over time, variations also result because each bird may copy elements of songs from more than one individual.

The amount of change in the song from generation to generation varies with each species of bird. White-crowned sparrows *(Zonotrichia leucophrys)* copy their species-specific song dialects extremely precisely (Baptista, 1975; DeWolfe and Baptista, 1995), whereas indigo buntings *(Passerina cyanea)* modify their song type slightly with each generation (Payne, 1996). Peter Slater and his colleagues (1980) have estimated that the chaffinch copies with an accuracy of 85 percent. In other species only the most common songs are sung from one year to the next. In some cases, what appear to be simple copying errors may occur because the songs heard are distorted by other noises in the bird's environment. L. Lehtonen (1983), who studied the songs of great tits *(Parus major)* in Finland, believes their songs have become simpler over recent decades because the environment has become more noisy. If environmental noises do affect song, we might well contemplate a world in which the songs of birds are degraded to their simplest form. A comparison of the urban members of a species with their conspecifics living in remote, wild environments might be an interesting way to test this hypothesis, but we would have to consider the potential influence of other factors that could also cause a difference in the songs sung by the two populations.

In fact, we know there is regional variation in the songs sung by birds of the same species. Birds living in one region may sing songs that are slightly different from those sung in a nearby region (see, e.g., Slater, 1986, 1989). The most common pattern is for the songs to change gradually as the distance between populations increases. This spatial variation in the song could be the result of a bird's copying different songs of more than one of its neighbors. Thus both time and spatial separation could contribute to changes in the songs. An alternative explanation, by B. B. DeWolfe and Luis Baptista (summarized in Bradbury and Vehrencamp, 1998), relates regional variation to migration; species that migrate and return to territories that may be some distance removed from the home range in which they learned their dialect might have to adjust their dialect to the new location. Sedentary species would have no need to do this and so would retain relatively stable dialects.

Great emphasis has been placed on the cultural transmission of vocalizations, but signaling in other sensory modalities may also be transmitted by learning. It now seems that birds may be able to learn visual signals as well as vocal signals. This would mean that both visual and vocal signaling could be passed on by cultural transmission.

Some patterns of signaling may also be used as a means to pass on information from one generation to the next—to assist cultural transmission. For example, European blackbirds use mobbing calls not only to attempt to drive a predator away but also to teach naive conspecifics that the predator is a threat to their survival (Curio, Ernst, and Vieth, 1978). In this way, young birds learn about predators from adults and the information is passed from one generation to the next.

VOCAL LEARNING IN NONPRIMATE MAMMALS

Compared with the many studies of vocalization in birds, there has been very little research investigating vocal learning in mammals (but see Janik and Slater, 1997). In particular, there have been fewer studies in which mammals have been experimented on by rearing them in isolation, compared to such studies in birds. This is probably because researchers have been much more aware of the ethical implications of raising mammals in conditions in which they are deprived of social contact than of raising birds in isolation, although this is an artificial distinction because birds are just as dependent on social relations as mammals. The hand-rearing experiments discussed above show that this is true.

To find evidence of vocal learning in mammals in the wild, we can first look to see whether any of them mimic sounds in their environment. Perhaps the best known example of a mammal doing so is that of Hoover, a harbor seal *(Phoca vitulina)* kept in the New England Aquarium in Boston. He learned to mimic human speech, including the phrases "Hello there" and "Come over here." Sound spectrograms of the seal saying these words and a human saying the same thing have been published by Katherine Ralls and her colleagues (1985) from the Smithsonian Institution (see Figure 6.3). The similarities between these spectrograms are

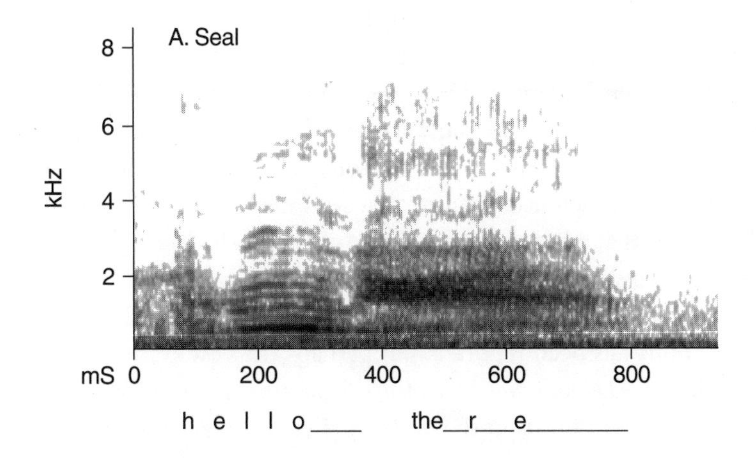

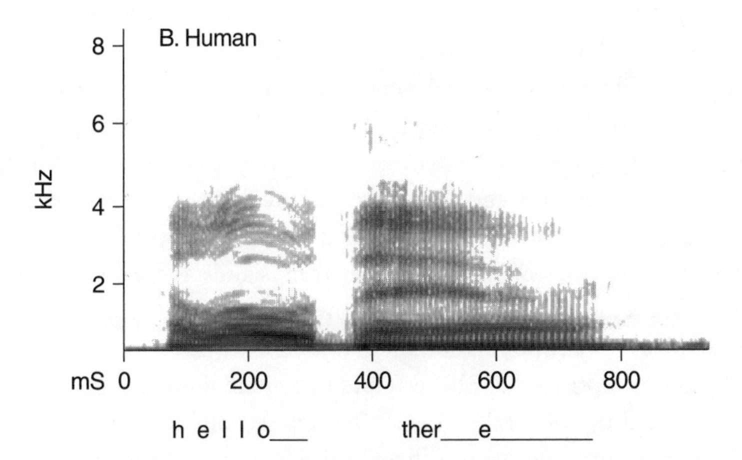

FIGURE 6.3 Mimicry of a human voice by a seal. A: The harbor seal *(Phoca vitulina)* called Hoover says "Hello there" with an American accent. B: The same words spoken by a human. (These sound spectrograms were made from an audiotape generously provided by Katherine Ralls, via James Scanlon. Other examples can be seen in Ralls, Fiorelli, and Gish, 1985.)

remarkable. Hoover also says, "Get out of there" and "Hey," which he strings into sequences with other sayings, and then ends with mimicry of human laughter. To the naked ear Hoover sounds like a human with an unnerving slur in the voice and a Boston accent. Another male seal in the same aquarium learned to say "Hello," showing that Hoover was not a unique case. Seals can learn human speech sounds when they are in circumstances that favor this type of learning. Hoover was reared without contact with his own species in early life.

It seems that vocal learning may not be uncommon in seals living in their natural environment. There is considerable variation in the vocalizations of members of the same species of seal living in different localities. Although there may be other reasons for spatial variations in vocalizations, these differences indicate that seals may learn their natural calls. It is possible that adult males mimic the calls of males in neighboring territories, as do some male songbirds.

There are several species of seals for which geographical variation in vocalizations has been reported, but perhaps the best example is that of Weddell seals *(Leptonychotes weddelli)*. Separate colonies of Weddell seals living in different fjords in the Vestfold Hills of Antarctica, only 20 kilometers apart, were found to share only 5 out of a total of 44 different vocalizations (Morrice, Burton, and Green, 1994). Even the shared calls were not absolutely identical. A much earlier study of elephant seals *(Mirounga angustirostris)* inhabiting islands off the west coast of North America, carried out by Burney Le Boeuf and Richard Petersen (1969), found that threat vocalizations made by males vary from one island's population to the next and that these local dialects have persisted from generation to generation. This variation occurred despite the fact that there was some movement of males between islands; the researchers suggested that young males that move to a new island copy the threat calls of the established male population on that island.

Whales in captivity have been found to imitate human speech, as was first reported by John Lilly (1965). Bottlenose dolphins *(Tursiops truncatis)* mimic their species-specific whistles and will also learn to mimic whistles that are used in training them to perform tricks. It appears that the dolphins' own whistles can be modified by experience. Diana Reiss and Brenda McCowan (1993) found that bottlenose dolphins mimicked computer-generated whistles and that they also learned to produce particular whistles in association with certain interactive behaviors, such as playing with rings or balls. There is no question that all the behavior of dolphins is highly plastic and creative, including their communication behavior. As far as we know, whales and seals are the most versatile vocal learners of all mammals.

VOCAL LEARNING IN PRIMATES

The other mammalian species in which vocal learning has been investigated to some degree are the primates. We have discussed the remarkable capacities of the great apes to learn sign language and symbolic forms of communication by which they can communicate with humans. This is ample evidence that they can learn complicated forms of communication. But do they learn their own vocalizations? Unfortunately, there have been surprisingly few studies of the learning of the species-typical calls of any of the apes. John Mitani (1994) and his colleagues have reported differences in the pant-hoot vocalizations (loud calls) of chimpanzees in two different localities in Africa, and they have some suggestive evidence that male chimpanzees calling at the same time match their pant-hoot sounds. This indicates that chimpanzee vocalizations can be shaped by learning, but more detailed investigations are needed. We summarize what is known of vocalizations made by orangutans in our book *The Orang-utans* (Kaplan and Rogers, 1999), but these data are patchy and no reliable developmental work has been done.

A study of contact calls in pygmy marmosets *(Cebuella pygmaea)* by Margaret Elowson and Charles Snowdon (1994) found that these monkeys modify their trill calls when their social environment is changed so that they can hear the calls of previously unfamiliar members of their own species. They modified both the frequency band width and the peak frequency of their trill calls, and these changes occurred in monkeys of all ages, from infants to adults. Thus the trill call, at least, of the pygmy marmoset is plastic, able to change, even in adults. Learning influences the call. Note the similarity to changes in birdsong. The ability to change vocalizations when the social environment changes may be essential to social cohesion in avian and mammalian species.

There is an important study by Robert Seyfarth and Dorothy Cheney

(1986) on vervet monkeys *(Cercopithecus aethiops)*. They managed to record eagle alarm calls made by 24 infants, 53 juveniles, and 55 adults (see Chapters 2 and 3) in the wild. They also noted the aerial species to which the eagle call was applied. Infants used the call to refer to birds flying overhead, but they did not call to all species of eagle and they often called when they caught sight of non-raptor, innocuous species, such as bee-eaters. Compared with the infants, juveniles showed a much greater awareness of different species of eagles, although non-raptor and innocuous species, such as storks, still incorrectly evoked their alarm calls. Adults, by contrast, gave the eagle alarm calls to refer to six different species of raptors, including the goshawk and the owl, and the only non-raptor that evoked their alarm calls was the vulture. However, the call for the vulture was observed to occur in fewer than five cases. The results show that infants use the eagle alarm call rather nonspecifically to refer to a wide range of aerial predators, whereas adults have learned to use the call specifically to refer to raptors. It took nearly 2 years to collect these data. They show very well that learning the meaning of a call must take place during growth from infancy to adulthood and that the monkeys may have particular images in mind when they produce this alarm call.

Marc Hauser (1988) has discovered that infant vervet monkeys also learn to recognize the alarm calls given by starlings. We have already discussed the vervet monkeys' attention to the alarm calls made by starlings that live in the same locality. The adult monkeys are able to exploit those calls to detect the presence of a predator in the air or on the ground. Given the interspecies nature of this form of signaling, it is not surprising that the vervet monkeys have to learn the meaning of the starlings' calls. Hauser conducted his research by playing back tape recordings of not only the starlings' ground-predator alarm calls but also their songs. Infant monkeys less than 3 months old were able to distinguish between the starlings' alarm calls and their songs, but they did not interpret the alarm calls as indicating danger. They simply looked at the loudspeaker when it was broadcasting a starling's alarm call but not when it was broadcasting the bird's song. By the time they were 3 to 4 months old, they had learned the meaning of the starlings' ground-predator alarm call—they responded to it by scampering up the nearest tree. Moreover, infants that had been exposed to more examples of the starlings' alarm call learned sooner than those exposed less often. These results need confirmation by research with more subjects, but they demonstrate interspecies learning of vocalizations. This is a very special case of learning.

Learning seems to occur for at least one other call made by vervet monkeys. Hauser (1989) studied the age-related changes in the *wrr* call that they make when they encounter another troop of vervet monkeys. He found that infants less than 3 months old produce a *wrr*-like call when they are distressed by being lost and that it is even more like the *wrr* call of adults than that of older infants. Although infants 10–18 months old produce *wrr* calls during encounters with other groups of monkeys, their calls are not acoustically identical to those of adults. In summary, very young infants (up to 3 months of age) make *wrr* calls when they are lost and at other times when they seek contact; then there is a period from 3 to 10 months of age when no *wrr* calls are made, followed by a period from 10 months to 3 or 4 years when *wrr* calls are made that are not the same as the adults' calls; and finally, the adult *wrr* is made after the monkey is 4 years old. This timetable for development of the calls was speeded up in infants belonging to groups that experienced more encounters with other groups of monkeys, a finding which suggests that learning plays a role in the development of the adult call used in a specific context. However, maturation of the vocal apparatus may also contribute to the development of calls. These transitions in the call that occur with age and/or experience are most interesting and deserve further study. Hauser noted that the period when *wrrs* were not produced (3 to 10 months of age) is also the period when other types of vocalizations are being acquired, and this other learning process might distract the young monkey from making the *wrr* call (see Figure 6.4). When it begins to make the call again, the structure of the call may have been degraded and that may be why it has to be learned again. Similar transitions have been reported in the development of language in human children.

Hauser (1994) has also shown, in a study of rhesus monkeys *(Macaca mulatta)*, that the characteristic dominant role of the left hemisphere in processing the species-specific vocalizations is not present in infants. It develops with increasing age. Hauser determined which hemisphere was dominant by scoring which ear the monkeys used to listen to a loudspeaker playing back their calls. Adults favored the right ear, and this means that most of the processing is occurring in the left hemisphere, because the auditory input goes mostly to the hemisphere on the side opposite the ear. Infants showed no preference for one ear over the other. These results do not tell us whether this developmental change in the processing of the vocalizations is influenced by experience, learning, or simply maturation, but it is possible that all these processes are involved. We know that experience influences the development of brain asymmetry in birds and learning can affect it, too, as the research of Lesley Rogers (1997a) has shown.

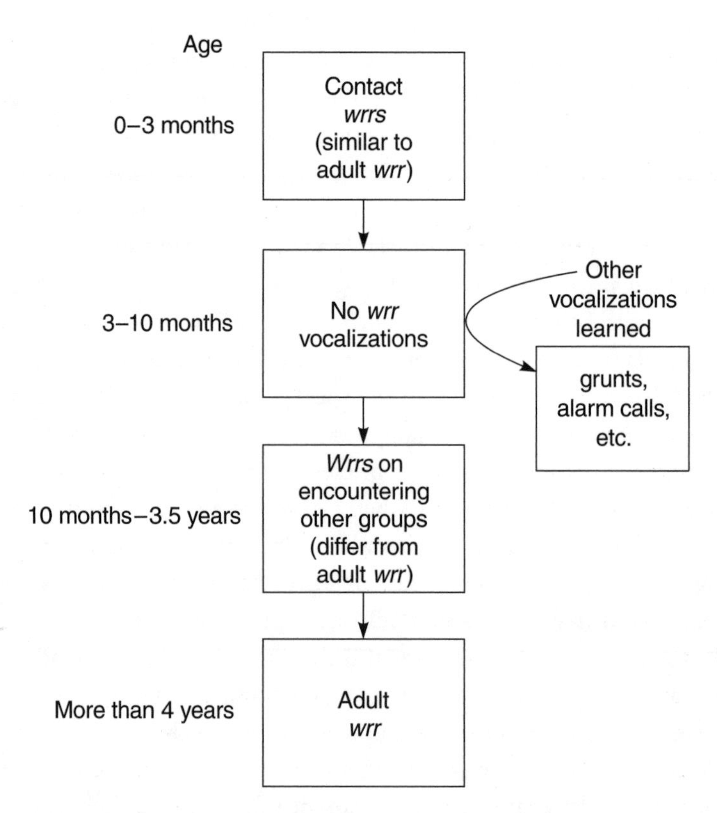

FIGURE 6.4 Developmental changes in the *wrr* calls of vervet monkeys. Adults produce the *wrr* call, a trilled call, during aggressive encounters between neighboring groups. Infants make a similar call when they become separated from their mothers. Note the period when no *wrr* calls are made, a stage when other calls are being learned. (After Hauser, 1989.)

Some studies on primates have found less evidence of vocal learning. The calls of the squirrel monkey *(Saimiri sciureus)* have been studied in some detail. Research carried out in the early 1980s by Anna Lieblich and her colleagues (1980) and by John Newman and David Symmes (1982) showed that there was little age-dependent change in the calls of squirrel monkeys, and also that the calls were not greatly affected when the monkeys were reared in social isolation from other members of their species. This result suggested that the calls of this species were strongly determined by inheritance—by the genes. However, more recent studies have indicated that this conclusion was incorrect. Maxine Biben and Deborah Bernhards (1995), of the National Institutes of Health in Maryland, have shown that the *chuck* calls of young females are more similar to those of members of their own social group than to those of members of other groups, a finding which suggests that learning is involved. Also there are marked differences in the types of calls that the monkeys make at different ages. Although this does not prove that learning occurs, it does show that there is flexibility in their vocalizations. There appears to be learning of the specificity of when and where these calls should be used.

Overall, it can be said that the present state of knowledge shows that marmosets, tamarins, squirrel monkeys, and many other primate species change their vocalizations with development, as John Newman (1995) of the National Institutes of Health has pointed out. In addition, we would like to return to a point that we made earlier: almost all the primate calls that have been studied are the more obvious calls, used to indicate alarm, distress, or contact with other members of the animal's social group; the more intimate and softer social calls have been ignored. It would be of great interest to compare the development of these calls with that of the more obvious calls.

CONCLUSION

From extensive research on song learning in birds, we can conclude that learning plays a major role during a sensitive period in early life. In some species, no new songs are learned in adulthood. In other species, such as the canary, adult song remains plastic, although seasonal changes in hormone levels determine when singing will occur.

It is simply too early to reach any firm conclusions about the learning of vocal communication in primates, or other mammals, but developmental changes do occur and at least some of these are known to be affected by experience and learning. There are some similarities between species of primates in the way the structure of their vocalizations changes with age. In addition, primates share with birds and humans the stage of vocal development referred to as babbling. Infant primates are very vocal compared with adults, and the highly varied subsong of songbirds is similar to the babbling of human babies. This would appear to be a stage when the young animal is practicing the structure of the vocalizations.

Not only must the structure of calls and song be learned but so too must the meaning of each call. The studies by Hauser were a step forward because they took into account the meaning of the call and did not just document the changes in the structure of the vocalizations with increasing age. There is enormous scope for more work along these lines.

It should also be noted that there is some evidence that different groups of monkeys belonging to the same species have different dialects (as Steven Green (1975) has shown to be true of Japanese macaques living in three different locations), and this too may indicate that learning of vocalizations occurs, as we explained for birdsong. With the vervet monkeys, where it is known that the meaning of calls is learned, it would now be interesting to know whether this type of learning forms a cultural tradition. In fact, we wonder whether the vervet monkeys of Barbados, taken there from Africa many years ago, have retained the same use of alarm calls and use them with the same specificity to refer to predators as do the vervet monkeys in Africa. Even if one generation teaches the calls to the next, the meaning may have changed to some extent with the presence of different predators in the new environment. Some information from the chimpanzees that have learned sign language shows that signs may be passed on from one generation to the next. Roger Fouts and his colleagues (1989) observed Washoe actively teaching her son, Loulis, how to sign: she was seen to mold his hands into the correct sign. This indicates that, in chimpanzees, the mechanisms for an active cultural tradition are in place.

Ryo Oda and Nobuo Masataka (1996) have studied the responses of two populations of ring-tailed lemurs *(Lemur catta)* to the different alarm calls that sifakas *(Propithecus verreauxi)*—another type of lemur—

give to aerial and ground predators. One population of the lemurs consisted of free-ranging groups living in the same area as sifakas, in Madagascar, and the other population was captive and had no contact with sifakas. The free-ranging group responded appropriately to the sifaka's calls, but the captive group was unable to tell the difference between the two calls. Thus experience with the sifaka's calls is important and learning must take place, but it is not known whether each individual learns the meaning of the two calls afresh in each generation or cultural transmission assists this process. The results of these studies show, however, that cultural transmission alone, without any relevant experience, does not suffice, because the captive group would have retained the ability to interpret the calls had that ability been passed down from previous generations.

Although most of the research on learning to communicate has focused on vocalizations, it should be remembered that primates, as well as other mammals, also communicate by scent marking, visual displays, and various forms of touch. It is important to investigate whether learning occurs in these other forms of communication. Is there, for example, a stage in development when visual displays are plastic, as there is in the development of song? It is not unreasonable to suggest that animals might show "babbling" in visual displays, as they do for vocalizations. In fact, Laura Petitto and Paula Marentette (1991) have shown that babbling in humans is not confined to vocal production: deaf children exposed to sign language from birth display manual babbling in which they run through the various hand positions used in signing, just as hearing children do by trying out sounds vocally. Thus an early phase of development when practice occurs is quite probably related to communication but is not necessarily confined to vocal communication. And since vocal babbling is not unique to humans, other forms of babbling may be found in animals too. As far as birds are concerned, we know much more about their vocal learning than we do about learning in any other group of animals, but we know virtually nothing of how they develop their visual displays, in all their complexity and variety.

Chapter Seven

THE EVOLUTION OF COMMUNICATION

In this chapter, we examine aspects of signaling and receiving that may be passed from generation to generation by a program encoded in the genes. Many important aspects of signaling are learned by each animal during the course of its life, particularly during its early life. But learning is not the only factor involved. All the behavior patterns that are used in communication depend on the interaction between genetic factors and experience or learning. Signals can be learned and passed on from generation to generation, as we saw in the discussion of the cultural transmission of song.

GENES AND LEARNING

In insects, some forms of simple behavior are programmed entirely by the genes, but the same is not true of the more complex behaviors used by vertebrates for communication. Almost all the forms of communication discussed in this book depend to various degrees on learning. But this does not mean that they have no genetic component. It is difficult, if not impossible, to look at the behavior of the signaler or the receiver and say how much of that behavior is determined by genes and how much depends on learning because the two interact and also because there is great variation from one animal to another. On this point we may differ from those biologists who consider genes to have a direct causal, and often overriding, influence on both signaling behavior and the processes involved in receiving and remembering signals. This genetic determinist view is encapsulated in those definitions of communication that require a signal to be adaptive, meaning genetically programmed. This is the definition of signaling formulated by Edward O. Wilson (1975) and held by John Krebs and Nick Davies (1993), as well as many other behavioral ecologists and also ethologists. Because they focus on the genetic program, they tend to be less interested in questions about learning and the development of communication behavior.

As we discussed in Chapter 6, songbirds reared in isolation sing abnormal songs but retain a template of the song. At first this result was interpreted to mean that genes determined the template and then learning elaborated upon that template to give the full song. In other words, each bird was said to inherit a foundation for its song, on which it builds by the process of learning that occurs when the young bird hears other birds singing. However, the experiments in which birds were rendered deaf early in life showed that even the template of the song is not present if the bird cannot hear itself sing. The results of such experiments, however offensive their methodology, do indicate that the earlier notion that genes alone determine the template for song may be incorrect, unless the effect of hearing no sound led to degradation of an existing template.

Nevertheless, genes do specify some broad aspects of the sensory systems that a particular species will have available to perceive signals—such as a sensory system for seeing certain wavelengths of light; ultrasound detection; the ability to detect electrical fields. This genetic endowment determines the nature of the signals that will be most effective for the species. Genes may also specify aspects of the structures that will be used to produce signals, such as the colors and sizes of feathers or the structure of the syrinx. In these cases, we may regard the genetic factors as constraints on the development of the structures and behaviors used for signaling, or as constraints on the development of sensory organs and the other processes that are used to detect, discriminate, and interpret the meaning of the signal. The amount of influence such genetic constraints exert on the development of the behavior of a species varies from species to species and from one form of communication to another, but in every case the genetic program provides only broad constraints. It does not determine the exact details of the behavior patterns used in communication.

Often, learning is not merely an elaboration, or fine-tuning, of a basic template that has been specified by the genes. In many cases, experience can so completely change behavior that the original genetic blueprint is no longer recognizable. The sensory capabilities of an animal depend on experience, as was clearly shown in experiments with kittens conducted by Colin Blakemore and Grahame Cooper. They discovered that a kitten that has been allowed to see only vertical black and white stripes for a short period after it first opens its eyes is unable to see horizontal stripes for the rest of its life (Blakemore and Cooper, 1970). The early experience so modifies the way the kitten processes visual information that, from then on, it does not respond to an object that is waved back and forth horizontally, whereas it runs to play with one that is moved up and down. The cells in the visual cortex that would normally respond to horizontal lines shift their preference and now respond to vertical lines. The opposite is true of a kitten exposed to horizontal stripes instead of vertical ones.

At first, this research might seem to indicate that genes make absolutely no contribution to the kitten's ability to see. However, a kitten that is kept in the dark over the same period of time as the others were exposed to the stripes is able to see both vertical and horizontal stripes, and indeed stripes at any angle. Therefore, the ability of the kitten to see stripes at any angle when it first opens its eyes appears to be determined genetically, unless another form of environmental stimulation acting before the eyes open also exerts some influence. Visual experience immediately after the eyes open has a critical effect in changing the way the visual system is wired, and it determines what the kitten will be able to see from that time on, what signals it will perceive. In the normal environment, a kitten would see lines at all angles and thus the early visual experience would reinforce the genetic plan. Only by putting the kitten in an abnormal visual environment can the importance of learning be shown.

We use this example of early experience in the kitten to illustrate the interaction of genes and early experience and, in this case, to demonstrate the overriding role of experience during a sensitive period of life. A similar effect of early experience has been shown by one of the authors (Rogers, 1995, 1996) to affect the development of the nerve connections that are used to process visual inputs in the chicken. Exposure of the chick embryo to light just before hatching stimulates the development of nerve cells that project from the chick's midbrain to its forebrain. In fact, because the embryo is oriented in the egg so that its head is turned with the left side against its body, only the right eye is stimulated by the light, which can pass through the eggshell and membranes that surround the embryo. Only the connections in the midbrain that receive input from the right eye are stimulated to grow. The connections on the other side of the brain that receive input from the left eye do not develop to the same extent because the left eye is not exposed to light. Therefore, an asymmetry develops in the visual pathways as a result of exposure to light before hatching. If chicks are hatched from eggs that have been incubated in the dark, no such asymmetry develops and there are fewer connections from both sides of the midbrain to the forebrain.

The visual behavior of the chick is also altered in ways that we would expect given the differences in the visual connections caused by light exposure or incubation in the dark. Chicks exposed to light before hatching can find and peck at grains of food scattered on a background of small pebbles when they are tested with a patch over the left eye but not when the patch is over the right eye. The chicks with a patch over the right eye display attack responses, whereas those with a patch over the left eye do so only rarely. Chicks hatched from eggs incubated in the dark do not show these asymmetries. They perform the same with both eyes. After incubation in the dark, chicks peck at grains and pebbles randomly. It is even possible to reverse the asymmetry by allowing the embryo's left eye to be exposed to light instead of the right eye. This is done by easing the embryo's head out of the egg just before hatching and putting a patch on the right eye. The left eye can then be stimulated by light. After this procedure has been carried out, the chicks hatch normally but have reversed asymmetry (Rogers, 1990).

The exposure of the developing embryo to light has these long-lasting effects on the visual connections to the forebrain and on the chick's behavior provided that the experience occurs just before hatching, when the visual connections are first becoming functional. In other words, there is a sensitive period during which light exposure has this effect. Exposure to light earlier during incubation or after hatching has no influence on the development of asymmetry in the visual connections or in behavior. This is a clear example of the way in which a specific kind of experience in early life can alter brain development and the way that information is perceived and processed subsequently. In fact, the asymmetry caused by light may also affect some aspects of social behavior and communication in the chick, since the eye used to look at another chick may signal whether it is likely to attack or not (left eye for attacking and right eye for not attacking).

Like this early visual experience, everything else an animal learns in early life can radically modify its subsequent behavior. The examples of song learning discussed previously illustrate this point. There are critical ages at which animals must be exposed to certain stimuli or at which they must learn certain things, and if these things do not happen, they will not develop in a way that is typical for the species. This principle applies to the behaviors used in communication as much as to any other behavior. Perhaps the most important thing to say here is that social learning in early life has many and various effects on the communication abilities and patterns that an animal develops.

It is, of course, essential for communication between members of the same species that all individuals share the same communication system, although there may be regional variations and also seasonal and age variations. Each individual's use of the common communication system can be established by social learning. It does not have to be programmed by genes, although sometimes it is assumed that the commonality of a particular signaling system implies that it is entirely genetically determined. Such a view may have its roots in considering animals as mechanistic and behavior as fixed action patterns. It denies the fact that behavior patterns can be learned consistently and well, with little variation between individuals or from one generation to the next.

We find it necessary to stress the role of learning because researchers interested in the evolution of displays and other forms of communication often tend to give it only passing recognition and then proceed to discuss evolution without any further mention of learning. We ask the reader to keep this in mind as we discuss the commonality of communication systems within and between species. When we speak of evolution, we are referring to the process of genetic selection. We do not deny that this process occurs and that it is particularly relevant to the physical structures that are used in signaling, such as feathers, organs used for sound production, and skin or eye color, but the signaling behavior itself may also be passed from generation to generation by learning and these two processes are intertwined, not separate. Unless the genetic and experiential contributions to a particular behavior pattern have been studied in detail, we will not follow the all-too-common practice of assuming that genes have the overriding role in determining the behavior pattern. We note that few evolutionary biologists would think that the entirety of the signaling behavior is determined by genes, but, in our view, they often place far too much emphasis on the genetic determinants at the expense of experience and learning. Genetic explanations for the complex behaviors that animals use to communicate tend to trivialize the processes of development.

THE EVOLUTION OF FEATURES TO ENHANCE SIGNALING

We have mentioned the specialized wing feathers that ducks preen during courtship displays. The behavior pattern of ritualized preening evolved along with the specialized structure of the feathers. The behavioral act of ritualized preening draws attention to these specialized feathers and the feathers, in turn, enhance the behavioral act itself.

The most striking example of feathers used for the purpose of display is the male peacock's train. The train has evolved to be so large that it is a considerable handicap to the bird in its day-to-day life, as we discussed in Chapter1. Yet this disadvantage is balanced by the train's effectiveness as a courtship signal. Raising the tail and bowing the head is a feature of courtship displays in other species of the pheasant family (including chickens and pheasants) and fanning of the tail accompanies these acts in other related species. The peacock courtship display is thought to have evolved from these simpler displays through exaggeration of the structure of the train in both its size and its visual attractiveness. The hundreds of eyespots visible when the male displays are attractive to humans as well as peahens.

There are other features that enhance displays. A colored beak makes displays with the beak more obvious, and a contrasting color in the skin, feathers, or fur around the eyes enhances any display using the eyes. Contrasting coloration around the eyes occurs quite commonly in birds and mammals. The color of the iris may also enhance displays in which the eyes are featured. The size of the pupil of the eye is more obvious if the iris is a light color. In fact, a dark-colored iris may be used to conceal the size of the pupil. It is interesting to note that the only way that humans can distinguish the sex of galahs *(Catacua roseicapella)* is by the color of the iris, males having a dark-brown iris and females a pink iris. The sex difference in the color of the iris means that males can detect the size of the female's pupil but females cannot so easily do likewise for the male's pupil. Signaling of an emotional state by pupil size might, therefore, in this species, be a female-to-male signal but not vice versa.

SEXUAL SELECTION

Since peahens select to mate with peacocks with more spots on the train of the tail, they may be the ones that have caused the evolution of the peacock's tail (Chapter 1). Whether they are really choosing on the basis of the number of eyes on the tail and not overall size or some other associated feature, such as brightness or color contrast, we cannot say, but they are selecting males with the biggest handicap. In so doing they may be choosing to mate with the healthiest peacocks and the ones with the best genes. Thus, female choice may lead to the evolution of bigger, larger, and brighter ornaments, such as the tail. This is known as the "good genes" hypothesis for explaining the evolution of physical characteristics that are used to signal. The hypothesis can be applied also to vocal signals in cases where females choose to mate with males that call more loudly and at faster rates. In some species of songbirds, larger song repertoires may be favored by females, and so the complexity of the singing patterns would increase (Nottebohm, 1972). Competition between males for territory and for priority access to females also leads to elaboration of songs and louder calling. This is considered to be an aspect of sexual selection since it depends on mating success.

Of course, any tendency to call more loudly or to have brighter and larger ornaments makes the male more conspicuous to predators, and a balance must be reached between attracting females and not becoming too conspicuous to predators (Ryan and Keddy-Hector, 1992). This is where the habitat of the species becomes important. In certain habitats the balance between sending the most conspicuous signal and remaining concealed from predators will be achieved in one way and in other habitats it will be achieved in another way, as we discussed earlier. As John Endler (1992) has said, a signal evolves as a local balance between the relative strengths of sexual selection and predation. If predation is the main factor influencing survival, color patterns have to be cryptic and vocalizations not easily detected or located. If there are few predation pressures, color patterns become more obvious, even garish, and vocalizations become louder and more easily located.

THE EVOLUTION OF SENSORY SYSTEMS AND PROCESSES USED TO PERCEIVE SIGNALS

Communication requires not only a signaler but also a receiver; evolutionary changes may occur on the receiver side of the dyad as well as on the sender side, as Tim Guilford and Marian Dawkins (1991) have pointed out. They discuss evolutionary changes in the sensory receptors used for detecting the signal as well as in the processes that are used to discriminate one signal from another and to decode or interpret the message that has been transmitted.

We have said that sexual selection depends on males advertising that they are physically fit or in good health, and that this requirement leads to the evolution of conspicuous ornaments and attractive calls. These signals might also be designed to stimulate the female's sensory system as much as possible. This idea is referred to as the "sensory exploitation" hypothesis. According to this hypothesis, the female's sensory system is driving evolution and the male adapts his signaling to fit her ability to perceive and respond to his signals.

It is possible that the female's sensory system is specialized to perform a function other than receiving the male's signals and that the male takes advantage of this specialization by adapting his signals to match this aspect of the female's perception. For example, if the sensory system of the female is designed for catching prey, the male will adapt his signaling to appeal to her way of finding prey (Ryan and Keddy-Hector, 1992). If the prey, such as an insect, is moving, the male may perform courtship displays that involve movement. If the prey is detected by color patterns, the male may use similar patterns in courtship displaying, and so on.

Research on the mating calls of túngara frogs belonging to the genus *Physalaemus* by Michael Ryan and his colleagues supports the hypothesis of sensory exploitation. The ability of the female frogs to perceive certain sound frequencies appears to be the process that drives the evolution of the calls by the male (Ryan et al., 1990; Ryan and Keddy-Hector, 1992). Males of the species *Physalaemus pustulosus* produce a whine-like call finished off by a "chuck" sound. The female's auditory system is designed to respond preferentially to the chuck part of the call. The females of a related species, *Physalaemus coloradorum,* prefer male calls with the added chuck sound over ones that are just a whine, even though males of their own species do not add the chuck to their call. This preference by the female can be determined by placing her in the center of an acoustic chamber and playing the male calls through speakers on opposite walls of the chamber. When one speaker plays a whine without the chuck and the other a whine with the chuck, the female approaches the speaker playing the whine with the added chuck.

Females of both species of frog have a preference for calls with the added chuck sound, but only the males of one of those species has managed to exploit that preference. This could mean that *Physalaemus coloradorum* males will eventually adapt their calling pattern to exploit their females' preference. Certainly, were these males to use the chuck, they would be preferred as mates over ones that do not use it. It is likely that males of a species ancestral to *P. pustulosus* and *P. coloradorum* did not add chucks to their calls and only *P. pustulosus* males evolved the ability to stimulate the female's sensory system to the best advantage. In other words, the female's sensory system appears to have had a preexisting bias that has been exploited by *P. pustulosus* males (Ryan and Rand, 1999). Such exploitation of the female's preference may come about by learning and cultural transmission within groups or populations, or by genetic selection. It is usually assumed that the signals of the male frogs are determined solely by genetic selection, but, so far, no experiments have been designed to see whether learning is involved.

REMEMBERING SIGNALS

Guilford and Dawkins (1991) postulated that the ability to make memories may evolve and so may be important in the evolution of signaling. The receiver often has to remember which animal sent the signal, whether the signal had been sent previously, and in what context it had been sent. Therefore, different capacities for memory could affect signaling. But could genetic selection improve the capacity for processing and remembering a specific signal? As we saw from the example of the kitten exposed to stripes, experience can have profound effects on sensory perception. Experience also affects signal interpretation and the memory processes involved in communication. Recognition of these radical effects of experience and learning leads us to believe that any hypothesis which considers only genetic selection of these abilities is likely to be too simplistic. We can speak of the evolution of the structure of the eye, for example, and consider how it changed over evolutionary time from one species to the next. But it is problematic to apply the concept of genetic evolution to the changes that might have taken place in the processing of visual memory, interpretation, and attention, since these are so malleable by experience. The same applies to other forms of sensory perception.

Of course, as the brain evolved, its capacity to process information and to store memories increased overall, but even a very simple brain can process, detect, decode, and remember quite complex signals. A more highly evolved brain may process and remember a greater number and range of signals because of its increased capacity, but it might not be able to process and remember any single signal better than a simple brain. This is another reason why we think that talking about genetic selection for an increased ability to process and memorize a single type of signal is problematic.

Discussions of genetic selection might better be confined to tangible elements of the sensory receptors that are used to detect the signal and perform some of the initial aspects of discriminating the signal from the background, and not be applied to memory processes. In the case of sensory receptors, single genes may influence a single factor, such as the presence of a particular visual pigment (so affecting color vision, which we discuss next). In such cases, it is not so difficult to make a link between genes and function. In vertebrates, at least, the more complex processes of decoding the signal and remembering that go on at higher levels of cognition are certainly not dependent on a single gene. Since these processes are heavily influenced by learning and experience, it would be an oversimplification to say that they are determined by genes in any unitary way.

Making memories depends on the expression of genes (for example, the "early" genes, known as c-fos and c-jun), changes in the connections between nerve cells, and changes in chemical and electrical transmission between nerve cells (Rose, 1992), but the making of any specific memory is not dependent on any single genetic characteristic. It is possible to block certain key processes in nerve cells and so affect an animal's ability to form memories. This can be done by injecting the animal with certain drugs that interfere with the specific cellular process being studied or by using genetic technology to target certain genes that make particular enzymes and so prevent them from being expressed. In fact, it is possible to "knock out" genes in specific regions of the brain, and the type of memory impaired by doing so can be determined. For example, mice with a genetic mutation that prevents formation of a particular enzyme (alpha-calcium-calmodulin kinase II) in the part of the brain known as the hippocampus have impaired spatial ability (Silva et al., 1992; Mayford et al., 1996). This enzyme, present in nerve cells, is essential for the electrical changes that occur in the hippocampus when spatial memories are formed. Manipulating the gene that enables the cells to make this enzyme has an effect on spatial memory. This result shows that genes are involved in memory formation and some of them have important roles. But although the mice may appear to have a rather specific form of memory loss when they are tested on one or two tasks in the laboratory, this is most unlikely to be the case if they were in their natural environment. An inability to remember spatial locations or use multiple spatial cues would affect a wide range of abilities essential for survival in the natural environment. Laboratory tests now use sophisticated molecular genetic techniques, but at the behavioral level the same tests may be often very basic. The methodology is therefore insufficient to prove that it is possible to genetically select animals that will have specialized abilities to make specific memories, such as memories that would be used in a particular form of species-specific communication.

When a memory is formed as a result of a learning experience, a cascade of cellular changes takes place (Rose, 1992). The activation of genes is part of this cascade, but the specificity of the cellular events and the memory itself have nothing to do with a single gene, or even a subset of genes, that are specifically related to forming that type of memory alone. Furthermore, experience and even the forming of memories itself affect the subsequent activation of genes. There are certainly genetic mutations that have an effect on learning and memory, but these are very nonspecific in their effects and so do not add empirical support to the hypothesis of Guilford and Dawkins about the selection of genes for making memories of specific signals. Perhaps, specific genes for specific memories will be found to exist in some forms of communication in invertebrates, but this is unlikely to happen in communication in vertebrates for the reasons we have outlined. But of course genetic selection may affect very broad capacities to learn and remember.

EYES AND EVOLUTION

There are also broad evolutionary contributions to the receiving of signals. Let us consider specializations of the eye in a few species of vertebrates. The eye of the frog is specialized to detect certain stimuli: there are cells in the retina that respond specifically to small spots, each about the size of a fly, as long as these spots are, or have been, moving. The retinal cells are called "bug detectors" (see the summary in McFarland, 1985). They have an obvious role in the feeding behavior of frogs and they may also be important in signaling behavior. It is as if the eye of the frog is a filter that allows the frog to attend to certain stimuli in preference to others. The same filter could be used for prey catching and signaling. Hence visual signals used by frogs would be attended to more actively if they involved the movement of small stimuli. In fact, two species of frog (*Staurois parvus* of Brunei and *Taudactylus eugellenis* of Australia) have been observed to signal during courtship by holding up an opened front hand, or foot, and waving it. Each of the frog's digits with their rounded ends would make a spot-like image on the receiver's retina and the waving would provide movement. This visual stimulus is very likely to stimulate the "bug detectors" in the retina. The visual signal has been matched to the perceptual capabilities of the frog, just as the vocalizations of the male túngara frog have been matched to the females' auditory system.

The retina of a bird's eye contains oil droplets of different colors located next to the cells that respond to stimulation by light, the photoreceptors. The oil droplets act as a kind of filter allowing the bird to attend more to some colored stimuli than to others. Exactly which color will be more attractive depends to some extent on the color, or colors, of the oil droplets a species has and also the visual pigments present in the photoreceptors. Chicks of domestic fowl *(Gallus gallus)* have pink oil droplets and they prefer to peck at red and yellow food grains. In addition to the filter in the retina, other processes in the brain may be involved in determining the red and yellow preference, but the oil droplets are thought to play a role. The preference for pecking at small red and yellow objects might have evolved because most of the grains that chickens eat in the wild are red to yellow in color. Once this color preference had evolved, it could have been applied to signaling behavior. Later in life, adult chickens develop red combs that signal their sex, state of health, and hormonal status. A preference for "seeing red" would enhance the signaling capacity of the comb. Thus the specialization of sensory perception, which might have evolved first for feeding, could be exploited for sexual and aggressive displays.

For further examples of the relationship between visual perception and visual signaling, let us consider color vision in more detail. An animal's ability to see color depends on the presence of color pigments in the receptor cells of the retina, and the presence of these is determined by genes. Humans have three such pigments (red, green, and blue) and they allow us to see the wavelengths of light spanning from red to violet. Because of these three pigments we are said to have trichromatic vision. Many other species of mammals have trichromatic vision also, but there are some that have only two color pigments and they are called dichromates. Interestingly, the marmoset *(Callithrix jacchus)*, the tamarin *(Saguinus fuscicollis)*, and the squirrel monkey *(Saimiri sciureus)*, all South American monkeys (called platyrrhine monkeys), are special cases in which all the males are dichromates together with some of the females, while the rest of the females are trichromates (see Jacobs, 1993). The reason for this sex difference in color vision is that some of the genes determining the pigments are carried on the X chromosome, the same chromosome that also determines an individual's sex.

The trichromatic females can see a greater range of colors than either the males or the dichromatic females, and consequently they probably have a better chance of finding ripe fruit in the dappled and changing light of the rainforest. But if trichromatic color vision is an advantage in finding fruit, why has it not conferred such a selective advantage on the trichromates that they have completely replaced the dichromates? Why have the dichromates not disappeared from existing populations of platyrrhine monkeys? It is possible that this evolutionary process is still in progress and that eventually the dichromates will be replaced by trichromates, but it is perhaps more likely that being a dichromate provides an individual with some other advantage that a trichromate lacks. Dichromates may be able to penetrate certain forms of camouflage and so detect prey that trichromates cannot. They may also be able to detect movement through foliage better than trichromates can. These abilities may be useful for finding foods other than ripe, colorful fruits, such as insects and nuts, which these monkeys also eat. Thus a mixed population of dichromates and trichromates would have a combined searching strategy superior to that of a single population of trichromates, and since these monkeys alert each other to the food they find, the combined knowledge would be shared.

The dichromatic and trichromatic forms of color vision may also confer different advantages in detecting different kinds of signaling. Marmosets, for example, use their tails as well as their faces in visual signaling. Their long tails have dark and light stripes that may be seen easily against a dappled background by dichromates, whereas the face has yellowish skin that changes hue in different states of arousal or sexual condition. This color change should be seen more clearly by trichromates. Thus groups of marmosets may consist of two types of individuals who pay different amounts of attention to different signals. This idea has yet to be tested.

Color vision is present in a large number of species. Most species of birds can see color better than we can, and they make full use of their color vision in displays using feathers in a rich variety of colors. Many avian species have four visual pigments in the retina: they are tetrachromates. It is known that several species of birds (pigeons, starlings, and zebra finches, for example) can see ultraviolet light—they can see shorter wavelengths of light than humans can. Indeed, it is likely that perception of ultraviolet light is widespread in birds, and Andrew Bennett and his colleagues have recently shown (1997) that the ultraviolet colors in the plumage of starlings and zebra finches are used in mating displays.

The only reptiles that we know to have been tested for color vision are two species of turtle, and they both have excellent color vision. It is more than likely that other reptiles can see color also, and we have already noted the use of color changes in displays by such reptiles as chameleon lizards. In fact, color vision evolved much earlier than reptiles, as a visit to a tropical coral reef makes eminently clear: there the fish are brightly colored and they use these colors in their displays. Bees also have color vision. Examination of those species that have color vision and those that do not has led researchers to conclude that color vision evolved separately several times over in different branches of evolution (Neumeyer, 1990). These separate appearances testify to the selective advantage that color vision confers on a species, although of course it is only an advantage in environments where color can be discriminated.

As we have already discussed, many rainforest birds are brightly colored and they use these colors to signal. The scarlet macaw *(Ara macao)* of the South American rainforest and the various birds-of-paradise that inhabit the rainforests of New Guinea, with their spectacular plumage, are the most striking examples of vibrant color in birds. In contrast to the wide variety of colors of the forest-dwelling birds, the colors of seabirds are more uniform, mainly black or brown and white. The reason is that the sea is a much more uniform environment than a forest and it is also very glary. Color is not easily distinguished against the glare of the sea or sky, and the ability to see color would not be highly advantageous in this environment. Seabirds may, however, find color a useful means of signaling at close range and where the amount of reflection is low. An excellent example is the yellow beak of the herring gull with its bright red spot, at which the gull chick pecks when the adult returns to the nest with a crop full of fish. The peck by the chick triggers the adult to regurgitate the fish and feed the chick.

Color vision is not useful to nocturnal species, so it was lost in nocturnal species. The earliest primates, the prosimians, are nocturnal and they have either very limited color vision or are completely color-blind, even though they evolved from species that, it appears, were able to see color. The primates that evolved later in evolutionary time may have "rediscovered" the color vision that their ancestors had lost, and they did so when they became diurnal and could benefit from having color vision (Mollon, 1990; Neumeyer, 1990). Alternatively, it could be argued that the extinct ancestor of both the prosimians and the higher primates was not, in fact, color-blind and that the loss of color vision in present-day prosimians is a more recent development.

Since the color vision of the diurnal primates of the Old World is trichromatic, it is thought to have evolved together with a food preference for colored fruits. Once it had evolved, color vision could be used for displays in primates. The displays of many diurnal primates depend on colors, whereas those of the nocturnal ones do not. The prosimians are mostly dull in color or have black and white stripes, as does the ring-tailed lemur *(Lemur catta)* on its tail. Among the later-evolving, diurnal species, the mandrill *(Mandrillus sphinx)* is the most striking exploiter of color, with a red and white striped snout and pink to blue skin on and around the genital area. Other diurnal primates, such as baboons and some macaque monkeys, have red, hairless skin on the buttocks that they present to other members of their troop as an appeasement display. The buttocks area of the female becomes redder when she is in estrus. This visual display, together with a change in the odor that she releases as a vaginal secretion, attracts the male and stimulates sexual behavior.

THE EVOLUTION OF VOCAL COMMUNICATION

So far, we have presented examples of evolutionary processes that may have been involved in visual displays, but apart from the discussion of the calls of the túngara frog, we have made only passing reference to the evolution of vocal communication and its relationship to auditory perception. We must emphasize that vocal signals are usually accompanied by visual signals, which are sometimes quite elaborate and are sometimes the postures that animals must adopt in order to produce the vocalization. In evolutionary terms, these two aspects of signaling are intimately linked. Unfortunately, researchers studying communication tend to concentrate on only one aspect of the signal pattern (usually either visual or vocal), and this inattention to other aspects limits our understanding of the entire signaling "package." This is particularly so in the study of vocal communication—the vocal signals have been described in great detail, but very little attention has been paid to the accompanying visual signals. We suggest two reasons for this. The first reason is related to the type of technology available for studying communication in the past: whereas it was relatively easy to take high-quality sound equipment into the field to record the vocalizations of animals, video records of animal postures in the field were, until recently, difficult to obtain because of cumbersome recording equipment and often impossible to obtain because researchers were unable to move the equipment to a place where they had direct sight of the subject.

The other reason for the focus on vocalizations has been the drive to understand the vocal communication of animals, particularly primates, in order to understand the evolution of human language. This focus on language has led many researchers to ignore anything other than the sounds made. If there has been any broader perspective than this, it has been to consider the gestures that primates make with their hands, and on the basis of such studies some researchers (e.g., Hewes, 1973) have considered the possibility that human language evolved from the gestures of primates. Communication by voice and communication by hands are human forms of communication. Little attention, however, has been paid to communication in primates by eye movements, body postures, odors, breathing patterns, or ear movements, any of which—alone or in combination—could have laid a basis for the evolution of human language.

It is not our aim to cover the evolution of human language here, but we remind the reader of the apes taught to use sign and symbolic language to communicate with humans and, in particular, of Kanzi's ability to understand the syntax of English. There are other characteristics of animals' ability to process sounds that are shared with humans and considered to be essential for language. These include the ability to control vocalizations and to use them referentially and the ability to perceive sounds categorically, as we discuss next.

Categorical perception is the ability to perceive sounds in categories that are discrete from one another, even though the variation in sound is actually continuous. The receiver picks parcels of auditory information out of this stream of variation and puts them into categories. Consider this example: Humans can hear the difference between "da" and "ta" sounds without any difficulty. If we use a computer to generate a range of sounds from "da" to "ta" so that there is a continuous gradation from one to the other and then play this range back to human subjects, they will say that they heard a collection of "das" and "tas" but not a continuum. Our perception creates a boundary between the two sounds that makes us believe there is a far more abrupt transition from "da" to "ta" than there actually is. We categorize the sounds, and this is said to be an important ability for understanding speech sounds. Not surprisingly, it was long thought that the ability to categorize sounds was uniquely human. We now know that that is not so. Brad May, David Moody, and William Stebbins (1989) have shown that Japanese macaque monkeys *(Macaca fuscata)* have categorical perception. The researchers selected two calls that the monkeys make when they want to establish contact with each other, one with a peak in frequency (pitch) early in the call and the other with a peak later in the call. From these they synthesized a range of calls, grading one into the other, and tested the monkeys with them to see whether they could distinguish one call from the other. The researchers found that although the monkeys were presented with a continuous gradation of calls, they perceived them as falling into two distinct categories, showing that the monkeys have categorical perception.

Categorical perception has also been demonstrated in chinchillas, which were tested with speech sounds; moreover, the boundary between one category and another found in chinchillas was the same as that in humans. Even Japanese quail categorize speech sounds, and a range of avian and primate species hear their own species calls categorically. There is now no question that this aspect of perception is shared by animals and humans (Kuhl, 1988).

The same is true of another aspect of vocal processing once thought to be unique to humans—lateralization, the processing of speech sounds and the production of speech by the left, not the right, hemisphere of the brain. For many years it had been thought that the specialization of the left hemisphere for speech and language processing was a characteristic unique to humans and a mark of our superiority to all other species. As John Bradshaw and Lesley Rogers (1993) have pointed out, there is now conclusive evidence that many species, including monkeys, mice, birds, and frogs, process or produce the vocalizations of their own species using only the left side of the brain. This attribute of the brain also evolved very early, contrary to beliefs once held.

Despite these similarities in auditory processing across many different species, there are, of course, differences between species in auditory perception and vocal production. These differences are determined by auditory experience, learning, and evolutionary processes. We will discuss some of these evolutionary differences briefly. We have seen already that bats can hear sounds that we do not and that they use these ultrasounds both to communicate and to signal. As in the case of vision, the hearing ranges of species vary and each species has evolved to match the transmission properties of the environment in which it lives.

A question often asked about the evolution of birds is, why have their songs become so complex? We discussed earlier the experiments in which John Krebs played back songs of European great tits in the field and found that the larger the song repertoire he played through the loudspeaker the more effectively birds were kept out of the area surrounding the speaker. The more complex the song, the better it is at advertising that the bird holds a territory. This is a plausible reason why more complex songs evolved. There is also some evidence that females prefer to mate with males with more complex songs, and this too would provide a reason for song complexity to evolve. It also provides a reason why the learning required to perform complex songs takes place.

Dialects of birdsong, regional variations among the members of one species, have been a source of speculation in regard to the evolution of song, but, so far, evidence is lacking that there is any link between song dialects and genetic differences. We will avoid further speculation on this topic and simply refer the reader to the book on birdsong by Clive Catchpole and Peter Slater (1995).

CONCLUSION

The evolution of communication is a topic that has attracted the attention of ethologists and anthropologists. There has been much speculation and some testable hypotheses. But at present, it is not a field in which we can establish many facts. This is partly due to the intangibility of evolutionary processes and the consequent difficulties they present for direct experimentation. We cannot go back in time to sample the potential effects of the genes of extinct species and, more particularly, we cannot observe the behavior of extinct species. Behavior does not leave a fossil record, but we can attempt to piece together the jigsaw of the evolution of communication by observing the signals and displays of existing species and design experiments to test hypotheses about these species. There are a multitude of exciting experiments lying ahead in this area of the study of communication, but we urge caution when researchers consider hypotheses that tie complex behavior and brain functions to unitary genetic causes, no matter how neat such links may appear to be.

Chapter Eight

HUMAN-ANIMAL CONTACTS

Human-animal communication occurs in many different contexts and takes a variety of forms. Without question, our attitude toward animals plays a significant role in the way we communicate with them, in the freedom we accord them, and in the manner in which we are willing to learn about their worlds and lives. Now, more than ever, we need to learn more about animals and more about our attitudes to animals. Our attitudes will ultimately decide how many species will have a future. Many human–animal encounters are not favorable for animals. Indeed, some contact with animals exists solely for the purpose of mass-producing them for consumption.

There is another side, however, the only one in which communication really plays a role, and that is contact between humans and animals as partners in work or as companions. These relationships can become very significant for us and possibly also for the animal. The dog, in particular, has been of great importance to humans for at least 12,000 years (Serpell, 1995). Animals bond with humans and many humans bond with their pets. Two to three thousand years ago, such a bond might have lasted for the best part of human life. In the days of the Roman Empire the human lifespan was about 24 years. Today, of course, humans live so much longer that pet owners tend to have serial relationships with dogs and cats. Domestication extended to goats and sheep about 9,000 years ago, followed by cattle and pigs and, in some areas, the horse (about 5,000 years ago).

The sheer passage of time makes us wonder how far domesticated species have changed as a result of becoming captives of human society and how this relationship might have affected their patterns of communication. The amount of change is likely to differ from species to species, partly because human contact varies among domesticated species and partly because, as Jonica Newby (1997) pointed out, most domesticated animals have never attained the status of closeness that dogs and cats have. Human-animal relationships have not remained static through the ages, yet their history remains largely unwritten. There are some notable exceptions, for instance the portraits of changing perspectives presented in works published by James Serpell (1995, 1996) and by Aubrey Manning and James Serpell (1994).

MYTHOLOGY AND FAIRY TALES

We begin with an unusual perspective on the human-animal relationship—namely, with fairy tales and popular mythology. Why raise them in a book on animal communication? First, fairy tales and myths about animals abound in all cultures and we might well read from them what a society desires or fears. In *Biophilia* (1984), for instance, Edward O. Wilson writes at length about the serpent as a mythical being that is feared in practice but held in awe in cultures around the globe. Yet other species are "poetic," as Wilson called them, because they arouse our curiosity. Second, in fairy tales, musicals, and fables, humans have always expressed their desire, if not their yearning, to understand what animals say and mean. Dr. Dolittle, for instance, thinks of animal communication as language that we have simply failed to learn but that we could learn. Third, animal communication is firmly embedded in the customs, folklore, and writings of many cultures. Finally, how fairy tales and mythology deal with animals—and herein lies our interest—may well mold the attitudes of young humans to animals and influence the kind of relationship with animals they will develop.

In contrast to the study of animal communication, which is fraught with difficulties in attributing reasonable explanations to acts of animal signaling, in the wonderful world of fairy tales stories unfold effortlessly and are clearly understandable. Animals of course feature in most fairy tales, but not always in the role of communicators. There are some stories, however, that focus our attention on animal communication. We cannot do more here than give just one example from a very rich field.

One of the most interesting stories we know in which human and animal worlds overlap is a fairy tale by Wilhelm Hauff called "Mutabor," written in the early nineteenth century. Hauff's tales are set in the era of Haroun-al-Raschid, the legendary ruler of Baghdad in the late eighth and early ninth centuries. We relate this story in some detail because it illustrates the assumptions of it makes about animals when speaking of animal communication.

In Hauff's book, a traveling group of businessmen tell each other stories as they slowly make their way on camel back through the desert. The story "Mutabor" is one of the most famous. In it, an evil magician wishes to get the sultan out of the way and, via a trader, offers the sultan a powder in a box with a Latin inscription. The inscription says that anyone sniffing this powder will turn into the animal he sees when turning toward Mecca and speaking the word "Mutabor." The sultan and his adviser promptly try this and turn into storks, since there are some storks in sight. Both men can now understand what the storks are saying. The inscription also says that one condition for their return to human form is that they must never laugh while they are in animal form. If they do, they will forget the word "Mutabor" and remain animals forever. However, on understanding what the storks are saying, both men are promptly very amused by a young female stork who is practicing a dance performance. They laugh heartily at what they perceive to be a clumsy attempt at dancing. They are now caught in stork form and find it difficult to adjust, particularly to the food, although they like their new ability to fly.

Eventually, the story leads us to a sad owl, shedding tears in a distant palace. She is actually a princess who was turned into an owl when she refused the hand of the evil magician's son. She can only be freed if someone proposes marriage to her while she is still an owl. The sultan/stork, a bachelor for many years, offers his hand in marriage, and she, still an owl, leads him and his adviser to the place where the magician always meets his supporters. There they hear the word "Mutabor" and can now change back into human form. The owl is transformed into a beautiful young girl and the sultan returns to his rightful place in society as a just and celebrated man.

There are several aspects of this story worth noting. First, natural curiosity drove the sultan and his adviser to sniff the powder despite the risks involved. We can assume that they were interested in knowing what animals had to say to each other. Second, they thoroughly enjoyed understanding the society of the stork. Third, we can assume that "stork language" translates into that of another animal species (here the owl), and that the sultan is a better man as a result of his experiences as an animal.

He has certainly been enriched, and the wife he could not find in human form he did find when he was an animal. On the other hand, capturing someone in animal form is obviously meant to be a punishment. The owl/girl cried because of the loneliness and the night she had to endure. The pivotal point of the story is how and when the owl and the storks would rediscover the magic word that would return them to human form. It was also made clear that the characters retained their identities as sultan and adviser, even as storks. Only the bodies had changed, not the minds.

One of the most telling parts of the story, and this is why we have related it here in detail, is that the evil magician predicted very accurately that both the sultan and his adviser would laugh once they were turned into animals. And of course the reader is equally caught up in the plot. Would we think that it is easy not to laugh? Or would we think, like the magician, that animal behavior is ridiculous or amusing and therefore that it would inevitably make us laugh? The story works only if it is assumed that laughter is inevitable.

The story also portrays the sultan and the adviser as acting in some ways like ethologists. Neither man actually enters the society of storks; they both just eavesdrop on their conversation. But, interestingly, by changing form they become "bilingual." They can still understand human language but they also understand the animals. By extrapolation this might also mean that animals can understand us even though we cannot understand them.

Respect for animals is probably a precondition for good communication with animals. In some human cultures respect for (some) animals is inbuilt as, for instance, in the Hindu religion, which considers cows sacred. In Australian Aboriginal cultures the conception of a new child is thought to be linked to an animal or a tree or a rock in situ. It is the presence of a lizard during conception that will give the child a lizard spirit. No doubt, such spiritual links can significantly alter the perception of animals by people holding beliefs of this kind.

HISTORIES OF HUMAN-ANIMAL RELATIONSHIPS

There are anecdotal stories and myths about animals that are alleged to have reared humans. Occasionally, we hear of so-called feral children who are supposed to have been reared by wild animals. If these children ever existed, they presumably learned to communicate with their foster families. These stories range from pure fiction to supposedly "real" cases. There is, for instance, the myth of Romulus and Remus, the founders of Rome, who were allegedly nourished by a shewolf. Then there is the story of a boy, Tarzan, who was described as growing up with apes. Other famous stories of feral children include that of Kaspar Hauser, that of the Wild Boy of Aveyron and, more recently, that of the Indian children Kamala and Amala, said to have been raised by a wolf (Maclean, 1977). There was also a boy reared by deer, who is said to have used his hands to imitate the movement of ears for communicative purposes. Douglas Candland's book on feral children (1993) provides fascinating examples.

Unfortunately, in modern cases in which children were thought to have been raised by animals, the reason for that assumption was that the human child or juvenile showed no mastery of human speech. The public interest these cases created was focused on documenting the development of the human capacity for speech rather than the communication and actual experiences of the person growing up in isolation from humans. But we may not have lost a golden opportunity. Steven Pinker (1994) from the Massachusetts Institute of Technology is right in suggesting that most of the stories are myths. In reality, these children were probably locked away in rooms by humans who deprived them of speech and communication, not reared by animals. (We mentioned the documented case of Genie in Chapter 6.) These cases, therefore, tell us little about animal-human contacts and communication.

Animal-human attitudes and communication stem from the history of specific animal relationships with human societies. Domestication is the most obvious case. Domestication has been defined by Juliet Clutton-Brock (1994) as a process by which an animal is bred in captivity for purposes of subsistence or profit, and lives in a human community that maintains complete control over its breeding, territory, and food supply. Clutton-Brock and other researchers in the field distinguish "domestication" from "taming," and this is a useful distinction in that taming may involve companionship. Taming may involve both humans and animals in work or leisure but may well exclude the final elements of domestication—targeting animals for slaughter. In both processes, the domestic or tamed animal is rightly regarded as a cultural artifact of human society. This is a very recent phenomenon in natural history.

In social histories, the domestication of animals has been treated as a watershed in human progress. Tim Ingold (1994) demonstrates that in many accounts the domestication of animals is perceived as signaling nothing less than the beginnings of "civilisedness." Perceptions of human progress are thus intrinsically tied to the subjugation of animals. Ingold argues that, from the nineteenth century until quite recently, only those who produced their own food were regarded as fully human. Hunting and gathering, or foraging, was considered not much better a way of life than the way of life of an animal, because hunting and gathering is precisely what animals do. The superiority and inferiority of human and animal societies could therefore be decided on methods of gathering food. Subsistence was inferior and surplus superior.

However, our views of hunter-gatherers have changed markedly in the last two to three decades, because more objective studies have revealed that hunter-gatherer societies were not "primitive." Nor are (or were) they wretched people at the point of starvation and lawlessness, as Charles Darwin thought. Indeed, some influential papers of the early 1970s showed that tribal societies were often affluent societies, without disease or immediate survival pressures. Moreover, the hunter-gatherers' knowledge of the behavior of animals, their sounds and their habits, was intimate. Only in this way could they hunt them successfully.

Domestication also required not only detailed knowledge of the species but a commitment, and often also resulted in an attachment. Pets are one example of such attachment, but people may even bond closely with working animals. For instance, cormorants are still used in some southeast Asian countries, chiefly Indonesia and Japan, to help their owners catch fish. Monkeys are used to pick coconuts, dogs for hunting game and foxes, and pigeons for relaying messages. In Mongolia to this day, deer are the most cherished (and often the only) possession of humans, and the people's relationship to the deer is embedded in mythology to the point that they associate their deer with their gods. Very often these ancient working relationships have been positive one-to-one relationships between humans and animals. The owners not only worked with the animals but slept with them. They shared food and their family life with them, often even allowing the animals to be unrestrained.

To get animals to do our work, there must be at least a degree of training and communication. Why else would animals, especially those that are much larger and stronger than humans, obey the whims of their owner or caretaker? In the last century in Thailand, elephants were taken into battle with neighboring countries, and today some elephants still work in the timber industry transporting logs. Mainly in Thailand, but also in Malaysia and India, there are still training schools for elephants. Each trainer spends a good deal of time with his elephant, and each elephant is teamed up with an older and more experienced elephant. Training may take up to 2 years. In that time the elephants learn to respond to verbal commands and requests. It is possible to achieve a level of cooperation that requires no threats of punishment and not even food rewards. In the forests of southern and southeast Asia, rider and elephant are often on their own. Trainer and trainee learn to trust each other through experience, consistency, and, not least, effective communication. Such communication cannot go just one way, from trainer to trainee. Some of it must also go the other way. The trainer must understand the ways of an elephant and must be able to read the elephant's signals appropriately. A trainer who does not understand when an elephant expresses anger or resentment may have a very short life.

The water buffalo plows the fields in Asia, still linked with its human master by the plow. Most of the peoples in the Middle East rely on the camel for personal transport and the transport of goods; they tend to care for them very well because ultimately their own lives are linked with those of their animals.

We do not want to romanticize relationships with animals that are built on their removal from a life of freedom with their own kind and that are, at times, subject to outright exploitation. What needs to be said, however, is that in these intimate forms of work relationships, the level of communication with the animal is often as good as or even better than that between pet and pet owner. Mutual reliance can occur only with good communication.

WAR ON ANIMALS

The Industrial Revolution in Europe probably did more to change our thinking about the nonhuman environment than any other single set of events. Production could be driven to new heights, and industrialization was based on the ideas of specialization and overproduction— the manufacture of surplus for profit. These attitudes were transferred to the nonhuman environment. Taking from nature whatever was thought necessary for sustaining a never-satisfied desire turned our relationship with the natural environment into one of indifference and alienation. Hunters may well have admired the strength, shape, or intelligence of animals they pursued, but people's attitudes toward animals changed with the development of industries that raised animals on a mass scale for the purpose of consuming them. Knowledge about animals and their behavior was no longer considered relevant unless it affected production. Nor was respect for animals and their world at all necessary in this kind of relationship between humans and animals. Such fundamental changes in attitudes can be traced readily in the imagery of travelogues, movies, and popular accounts over the last 150 years. It is noticeable, as we found when we researched the imagery of orangutans in such source materials, that modern attitudes vacillated between fear and indifference, but rarely included respect (Kaplan and Rogers, 1995). The Industrial Revolution bequeathed a remarkable legacy. Not only do goods and services carry a price tag in human society, but the natural world does too. Every plot of land, every tree, every ecosystem, every animal has been assigned a price at one time or another in the twentieth century. And once nature was "priced" in this fashion, it has been difficult to maintain a balance between greed and sustainability. The extinction of species of both flora and fauna is now a daily event and the number of species lost if growing exponentially.

In the second half of the twentieth century, we have encountered additional ethical problems as genetic engineering pushed arguments about the rights of animals to new limits, as Colin Tudge (1993) pointed out a few years ago. Do animals have the right to remain untampered with genetically? "Designer" animals are bred to provide "spare parts" of living tissue for humans. The image of the mouse carrying an unfurred human-shaped ear on its back may be bizarre, but it symbolizes a new, and this time grotesquely visible, peak in animal exploitation. Cloned animals are now also a reality. Needless to say, historically these new activities represent the lowest point in human–animal relationships—and also the lowest point in human-animal communication.

Simultaneously, however, there have been strong counterforces, concerned with the ethical issues of raising animals for profit on the one hand and the conservation issues of wildlife on the other. Even mainstream thinking in Western societies had to admit that our management of the natural world has led to an impoverishment of our understanding of the natural world (Nelson, 1987; Kellert, 1994). Now there are conservation societies, wildlife protection societies, ethics committees, animal liberation societies, and a myriad other projects spawned by alarm at the treatment of nature and animals. The Great Ape Project, for instance, advocates the rights of apes as our closest evolutionary relatives. Animal protection societies and societies for the prevention of cruelty to animals have impressively large numbers of members and they are very numerous.

Yet for many species, the slide toward extinction continues. This is partly so because certain species offer "products" for which some people will pay high prices, such as the ivory of elephants, the skin and teeth of tigers, the horn of rhinoceroses, the live bodies of apes for pets, zoos, and circuses, the whale meat for speciality restaurants and many other species, such as birds, for our amusement. Other species, such as the orangutan, could be marked to vanish from this earth because of the space for agriculture, mining, forest timber, and even habitation that we humans take from their natural habitat. The national parks and wildlife reserves that have been steadily created throughout the twentieth century have often been isolated areas. They usually occupy less than 1 percent of the area within the political boundary of a nation, too small in the long run to sustain diversity and too small to sustain healthy populations. Many species, such as bald eagles, peregrine falcons, ospreys, and some owls, have been severely damaged by pesticides, and only the banning of DDT and other dramatic interventions brought some species back from the brink of extinction. Australia, like many other countries in the world, is at risk of losing almost all its owl species (in Australia, owls are all on the vulnerable or endangered list).

Only a few species have genuinely benefited from expanding human habitation, and these are mostly species that are not popular, such as invertebrates (especially cockroaches), rodents (mice and rats), and opossums and squirrels. Among bird species, there are the sparrows, the crows, and the pigeons. Many of these are treated as vermin. In cities, especially in industrialized countries, human tolerance for animals, including insects, in close proximity has often dramatically declined. Humans routinely buy household sprays and rodent poisons. Pesticide spraying is often higher in cities than in rural areas. In homes, gardens, streets, and public spaces, the human inhabitants have declared war on all the remnants of living things. The very same people may then keep pets. This seems only to be a contradiction. What it indicates is that in modern urban areas, in particular, humans expect to have total control over animals and to make decisions about what species can coexist with them at any time. Part of the history of human-animal relationships is the dismal story of the destruction of human-animal coexistence. Inevitably, such developments in human history have brought with them highly selective perceptions of animals and attitudes toward their welfare, let alone attitudes toward communicating with them.

POSITIVE BONDS AND THE BENEFITS OF ANIMALS

Despite the bleak picture of ecological crises that can be drawn at the beginning of the twenty-first century, there remain attempts to try to coexist with and to see value in animals beyond profit or other selfish uses. In laboratories, in research stations, and on some farms, there are attempts to provide animals with better living conditions. Some people have started to acknowledge that animals have preferences, interests, and needs beyond physical survival. Marian Dawkins (1993) argues for giving animals choices in the selection of their environment. Why not ask them, she proposed? Of course, doing this entails not only respect for animals, but also the realization that animals can tell us something—that they can effectively communicate with us.

Many people keep pets and here communication between human and animal may work relatively well because of the species selected as pets. First, we tend to choose species with communication systems that operate largely within our own range of perception. We may not hear as well as dogs or see as well as cats in the dark, and we do not have the same color perception as birds, but we can hear them, see them, touch them, and speak to them. And, second, when we do get close to our pets, we usually claim to understand them. Our pets train us as much as we are supposed to train them. We learn to respond to their wishes to leave the house. We seem to be perfectly aware when they want to be near us, want affection, attention, or simply their food. We can understand some of their signals and they understand ours.

Beyond humans' desire to share their private lives with pets, many industrial societies have now also designed programs in which animals are used for more benign purposes than in the past. Boris Levinson (1969) discovered the great advantages of companion animals in clinical psychology and medicine. Research has since been done showing that animals can reduce stress in humans, that companion animals can increase self-esteem, and, even more dramatically, that they can lower levels of accepted risk factors for cardiovascular disease (Blackshaw, 1996).

Interestingly, these "uses" of animals are often related to communication and the senses. This is a relatively new field. A recent study showed that dogs can identify matching human body scents 80 percent of the time. A dog's olfactory sensitivity, selectivity, and memory, as well as its capacity for odor pattern recognition, is used in criminal investigations and security operations. Ray Settle and his colleagues (1994) predict that these skills are unlikely to be challenged by any artificial sensor in the foreseeable future.

The literature today also refers to a "human–companion animal bond." There are now programs in place throughout the Western world that have placed the human-animal bond on a new footing. Pet Facilitated Psychotherapy is one such program. Better known are the Pets as Therapy programs and Seeing Eye Dogs for the Blind. There are also Hearing Dogs for the Deaf. A study by G. Guttman, M. Pedrovic, and M. Zemanek in 1985 found that children who have pets not only have greater self-esteem than those without pets but are also better in nonverbal communication. Lynette and Benjamin Hart and B. Bergin (1987) argue that people in wheelchairs who participate in any of the health programs with pets tend to smile more, are greeted more often, and engage in conversation to a much greater extent than wheelchaired people without pets. Pets are thus regarded as great facilitators in communication among humans, and there are clear benefits for humans in keeping pets, whether for services they perform (cats killing vermin or dogs protecting the home) or companionship they provide.

Relatively little has been written about the actual communication between animals and humans, or the quality of life that animals are afforded by humans. In an industrial society that overall appears anti-animal, it is often difficult for people to develop attitudes that give primacy to the interests of animals, let alone develop the willingness to communicate with animals—close and specific domestic bonds excepted. On many occasions, humans seeking contact with animals may not be aware that this is not welcomed by the animals and that not all animals feel privileged to be singled out for human attention. In today's world, most wild animals are afraid of humans. They have cause to be. For those of us who rehabilitate wildlife there is a further lesson, sometimes quite shocking, to be learned: Most of the time, those animals do not need or want us at all. The best service we can do the ones remaining in the wild is to leave them alone and let them remain in a habitat in which they can thrive, if we can still do that. Arnold Arluke (1994) describes these issues as the modern contradictions in the relationship between animals and humans, who shower some animals with affection while simultaneously maltreating and killing other animals. Ironically, at the very time in history when we have probably the most widespread association with pets, and loving or romantic bonds with some animals are at a new historical peak, countless species in the wild are slipping quietly into extinction.

CONCLUSION

Darwin, and other authors in the twentieth century, have bequeathed to us several substantial problems in our relationship with animals, by telling us that animals in some way or another are our evolutionary ancestors. This notion compelled many to fight hard against the idea of continuity with animals. At the same time the Darwinian theory of evolution has remained a plausible scientific explanation for the development of life on earth. Within that theory, however, there continued to be an emphasis on upholding the uniquenesses of human beings as the pinnacle of creation. This distinction between humans and other animals has begun to be broken down by studies showing that nonhumans have brain asymmetry, the ability to use tools, the capacity to solve problems, sensory perception, the ability to learn, and a host of other capacities that were previously considered restricted to humans.

The communication system is one of the chief systems considered to distinguish humans from animals. Controversy has raged throughout the twentieth century about the possibility that animals have a language-like system of communication, and about the possibility that animals can communicate effectively with humans in human language terms. Herbert Roitblat, Heidi Harley, and David Helweg (1993) rightly point out that little work in psychology has engendered as much emotional involvement and heated argument as research into animal "language."

Throughout this book we have referred to research that has put its energy into training dolphins, apes, and birds to acquire sufficient communication skills (either as vocalizations or as symbols or sign language) to communicate across the species divide. In this effort, researchers have tried to learn more about the possibility of consciousness in animals, their ability to make use of past events or ponder future events and choices. These training and communication efforts have shown extreme dedication on the part of individual humans. They have often involved a lifetime of work, and a gracious indulgence on the part of the trained animal, which, after all, was entirely deprived of any of its natural life alternatives and often also of same-species companionship. Together humans and animals have lived and grown to explore the possibilities of meaningful communication across human/animal borders and to answer perhaps some of the questions about the extent of similarities and differences between humans and some animals. Humans' interest in such work has many sources and probably many intellectual justifications. One of the most relevant is the desire to learn more about the evolution of linguistic competencies. We cannot study this from fossil records because communication, for all its vitality and importance, leaves no trace in fossils.

We have referred equally in this book to studies that have placed animals in close proximity to humans (the laboratory or even the home) and to those that have studied animal communication in the natural setting. Charles Snowdon (1993) has called ethologists "cross-species anthropologists." It is only in the natural environment that answers to a number of questions concerned with communication will be forthcoming. For instance, as Snowdon says, ethologists may do studies in the field designed to answer questions about the evolutionary precursors of various linguistic phenomena: What are the environmental conditions that might have led to symbolic communication? What are the circumstances that lead to syntactic structures? What developmental influences affect the acquisition of phonology, comprehension, or usage?

It needs to be asked whether an emphasis on vocal and hence linguistic development is adequate. As we have seen, communication systems are complex and may involve several senses at once, some of which the human observer is capable of studying only by developing technological aids for their detection. Equipped with the naked eye or ear and our sense of smell alone, we would never have discovered the diversity and complexity of animal communication that we now know about, and surely much remains to be discovered.

Researchers from very different fields and with very different agendas might well agree that the study of animal communication, and the consequent discovery of commonalities of some aspects of communication across species, raise the possibility of viewing human language as one of several alternative systems of communication. Never before in human history has there been such intense engagement with animals in a scientific manner to try to understand how they communicate with each other and how they may communicate with humans. Never before has there been such an urgent need to undertake such studies.

REFERENCES

INDEX

REFERENCES

Adams, E. S., and Caldwell, R. L. 1990. Deceptive communication in asymmetric fights of the stomatopod crustacean *Gonadodactylus bredini. Animal Behaviour,* 39, 706–717.

Adret, P. 1993. Operant conditioning, song learning, and imprinting to taped song in the zebra finch. *Animal Behaviour,* 46, 149–159.

———. 1997. Discrimination of video images by zebra finches *(Taeniopygia guttata):* direct evidence from song performance. *Journal of Comparative Psychology,* 111(2), 115–125.

Altringham, J. D. 1996. *Bats: Biology and Behaviour.* Oxford University Press, Oxford.

Andrew, R. J. 1956. Some remarks on behaviour in conflict situations, with special reference to *Emberiza* spp. *British Journal of Animal Behaviour,* 4, 41–45.

———. 1961. The displays given by passerines in courtship and reproductive fighting: a review. *Ibis,* 103, 549–579.

———. 1965. The origins of facial expressions. *Scientific American,* October, 88–94.

———. 1972. The information potentially available in mammal displays. In R. A. Hinde (ed.), *Non-verbal Communication.* Cambridge University Press, Cambridge.

Arluke, A. 1994. Managing emotions in an animal shelter. In A. Manning and J. Serpell (eds.), *Animals and Human Society: Changing Perspectives,* pp. 145–165. Routledge, London.

Baptista, L. F. 1975. Song dialects and demes in sedentary populations of White-crowned Sparrow *(Zonotrichia leucophrys nuttalli). University of California Publications in Zoology,* 105, 1–52.

Barrett-Lennard, L. G., Ford, J. K. B., and Heise, K. A. 1996. The mixed blessing of echolocation: differences in sonar use by fish-eating and mammal-eating killer whales. *Animal Behaviour,* 51(3), 553–565.

BBC Worldwide. 1996. *Parrots: Look Who's Talking.* Video. Natural History Unit, London

Bennett, A. T. D., Cuthill, I. C., Partridge, J. C., and Lunau, K. 1997. Ultraviolet plumage colors predict mate preferences in starlings. *Proceedings of the National Academy of Sciences,* 94, 8618–8621.

Biben, M., and Bernhards, D. 1995. Vocal ontogeny of the squirrel monkey, *Saimiri boliviensis peruviensis.* In E. Zimmermen, J. D. Newman, and U. Jürgens (eds.),

Current Topics in Primate Vocal Communication, pp. 99–140. Plenum Press, New York.

Blackshaw, J. K. 1996 Developments in the study of human-animal relationships. *Applied Animal Behaviour Science,* 47 (special issue on human-animal relationships), 1–6.

Blaich, C., Steury, K. R., Pettengill, P., Mahoney, K. T., and Guha, A. 1996. Temporal patterns of contact call interactions in paired and unpaired domestic zebra finches *(Taeniopygia guttata).* Paper delivered at the International Society for Comparative Psychology Meeting, Montreal.

Blakemore, C., and Cooper, G. F. 1970. Development of the brain depends on the visual environment. *Nature,* 228, 477–478.

Blest, A. D. 1957. The function of eyespot patterns in the lepidoptera. *Behaviour,* 11, 209–256.

Blumstein, D. T., and Armitage, K. B. 1997. Alarm calling in yellow-bellied marmots: I. The meaning of situationally variable alarm calls. *Animal Behaviour,* 53, 143–171.

Boinski, S. 1991. The coordination of spatial position: a field study of the vocal behaviour of adult female squirrel monkeys. *Animal Behaviour,* 41(1), 89–102.

Borgia, G. 1995. Complex male display and female choice in the spotted bowerbird: specialized functions for different bower decorations. *Animal Behaviour,* 49(5), 1291–1301.

Bradbury, J. W., and Vehrencamp, S. L. 1998. *Principles of Animal Communication.* Sinauer, Sutherland, Mass.

Bradshaw, J. L., and Rogers, L. J. 1993. *The Evolution of Lateral Asymmetries: Language, Tool Use, and Intellect.* Academic Press, San Diego.

Brindley, E. L. 1991. Response of European robins to playback of song: neighbour recognition and overlapping. *Animal Behaviour,* 41, 503–512.

Brown, C., Gomez, R., and Waser, P. M. 1995. Old World monkey vocalizations: adaptation to the local habitat? *Animal Behaviour,* 50(4), 945–961.

Bullock, T. H., and Heiligenberg, W. 1986. *Electroreception.* Wiley & Sons, New York.

Busnel, R.-G. 1977. Acoustic communication. In T. A. Sebeok (ed.), *How Animals Communicate,* pp. 233–252. Indiana University Press, Bloomington.

Byrne, R. 1995. *The Thinking Ape: The Evolutionary Origins of Intelligence.* Oxford University Press, Oxford.

Caldwell, M. C., and Caldwell, D. K. 1965. Individualized whistle contours in bottlenosed dolphins *(Tursiops truncatus). Nature,* 207, 434–435.

Candland, D. K. 1993. *Feral Children and Clever Animals: Reflections on Human Nature.* Oxford University Press, New York.

Caro, T. M. 1995. Pursuit-deterrence revisited. *Trends in Ecology and Evolution,* 10, 500–503.

Catchpole, C. K., and Slater, P. J. B. 1995. *Bird Song: Biological Themes and Variations.* Cambridge University Press, Cambridge.

Catchpole, C. K., Dittami, J., and Leisler, B. 1984. Differential responses to male song repertoires in female song birds implanted with oestradiol. *Nature,* 312, 563–564.

Cheney, D. L., and Seyfarth, R. M. 1985. Vervet monkey alarm calls: manipulation through shared information? *Behaviour,* 94, 150–166.

———. 1990. *How Monkeys See the World: Inside the Mind of Another Species.* University of Chicago Press, Chicago.

Chevalier-Skolnikoff, S. 1973. Facial expression of emotion in nonhuman primates. In P. Ekman (ed.), *Darwin and Facial Expression,* pp. 11–90. Academic Press, New York.

Clark, L. and Mason, J. R. 1987. Olfactory discrimination of plant volatiles by the European starling. *Animal Behaviour,* 35, 227–235.

Clutton-Brock, J. 1994. The unnatural world: behavioural aspects of humans and animals in the process of domestication. In A. Manning and J. Serpell (eds.), *Animals and Human Society: Changing Perspectives,* pp. 23–35. Routledge, London.

Clutton-Brock, T. H. and Albon, S. D. 1979. The roaring of red deer and the evolution of honest advertisement. *Behaviour,* 69, 145–170.

Cohen, S. 1994. *The Intelligence of Dogs. Canine Consciousness and Capabilities.* Free Press/Macmillan, Toronto.

Crook, J. H. 1969. Function and ecological aspects of vocalisation in weaver birds. In R. A. Hinde (ed.), *Bird Vocalisations,* pp. 265–289. Cambridge University Press, Cambridge.

Cruickshank, A. J., Gautier, J.-P., and Chappuis, C. 1993. Vocal mimicry in wild African Grey Parrots *Psittacus erithacus. Ibis,* 135, 293–299.

Cullen, J. M. 1972. Some principles of animal communication. In R. A. Hinde (ed.), *Non-verbal Communication,* pp. 101–122. Cambridge University Press, Cambridge.

Curio, E., Ernst, U., and Vieth, W. 1978. The adaptive significance of avian mobbing. II. Cultural transmission of enemy recognition in blackbirds: Effectiveness and some constraints. *Zeitschrift für Tierpsychologie,* 48, 184–202.

Curtiss, S., Fromkin, V., Krashen, S., Rigler, D., and Rigler, M. 1974. The linguistic development of Genie. *Language,* 50, 528–554.

Davies, N. B., and Halliday, T. R. 1978. Deep croaks and fighting assessment in toads *Bufo bufo. Nature,* 274, 683–685.

Dawkins, M. S. 1986. *Unravelling Animal Behaviour.* Longman, Harlow, U.K. (2nd ed., 1995.)

———. 1993a. Are there general principles of signal design? *Philosophical Transactions of the Royal Society of London B,* 340, 251–255.

1993b. *Through Our Eyes Only? The Search for Animal Consciousness.* W. H. Freeman Spektrum, Oxford.

de Luce, J., and Wilder, H. T. (eds.). 1983. *Language in Primates: Perspectives and Implications.* Springer-Verlag, New York.

Dowsett-Lemaire, F. 1979. The imitative range of the song of the marsh warbler

Acrocephalus palustris, with special reference to imitations of African birds. *Ibis,* 121, 453–468.

Eibl-Eibesfeldt, I. 1972. Similarities and differences between cultures in expressive movements. In R. A. Hinde (ed.), *Non-verbal Communication,* pp. 297–314. Cambridge University Press, Cambridge.

Ekman, P. (ed.). 1974. *Darwin and Facial Expression.* Academic Press, New York.

Elowson, A. M., and Snowdon, C. T. 1994. Pygmy marmosets, *Cebuella pygmaea,* modify vocal structure in response to changed social environment. *Animal Behaviour,* 47, 1267–1277.

Endler, J. A. 1992. Signals, signal conditions, and the direction of evolution. *The American Naturalist,* 139, S125–S153.

———. 1993. The color of light in forests and its implications. *Ecological Monographs,* 63(1), 1–27.

Epple, G. 1968. Comparative studies on vocalization in marmoset monkeys *(Hapalidae). Folia Primatologica,* 8, 1–40.

———. 1975. The behavior of marmoset monkeys *(Callithricidae).* In L. A. Rosenblum (ed.), *Primate Behavior: Developments in Field and Laboratory Research,* pp. 195–239. Academic Press, New York.

Epple, G., Küderling, I., and Belcher, A. 1988. Some communicatory functions of scent marking in the cotton-top tamarin *(Saguinus oedipus oedipus). Journal of Chemical Ecology,* 14(2), 503–515.

Evans, C. S. 1997. Referential signals. *Perspectives in Ethology,* 12, 99–143.

Evans, C. S., and Marler, P. 1991. On the use of video images as social stimuli in birds: audience effects on alarm calling. *Animal Behaviour,* 41, 17–26.

———. 1994. Food calling and audience effects in male chickens, *Gallus gallus:* their relationships to food availability, courtship and social facilitation. *Animal Behaviour,* 47, 1159–1170.

Evans, C. S., Evans, L., and Marler, P. 1993. On the meaning of alarm calls: functional reference in an avian vocal system. *Animal Behaviour,* 46, 23–38.

Fadem, B. H. 1985. Chemical communication in gray short-tailed opossums *(Monodelphis domestica)* with comparisons to other marsupials and with reference to monotremes. In D. Duvall, D. Müller-Schwarze, and R. M. Silberstein (eds.), *Chemical Signals in Vertebrates,* vol. 4: *Ecology, Evolution, and Comparative Biology,* pp. 587–607. Plenum Press, New York.

Farabaugh, S. M., Linzenbold, A., and Dooling, R. J. 1994. Vocal plasticity in budgerigars *(Melopsittacus undulatus):* evidence for social factors in the learning of contact calls. *Journal of Comparative Psychology,* 108, 81–92.

Feduccia, A. 1996. *The Origin and Evolution of Birds.* Yale University Press, New Haven.

Fenton, M. B. 1994. Assessing signal variability and reliability: "to thine ownself be true." *Animal Behaviour,* 47(4), 757–764.

Fernald, A. 1992. Human maternal vocalizations to infants as biologically relevant signals: an evolutionary perspective. In J. H. Barkow, L. Cosmides, and J. Tooby (eds.), *The Adapted Mind: Evolutionary Psychology and the Generation of Culture,* pp. 391–428. Oxford University Press, New York.

Ficken, M. S., Hailman, E. D., and Hailman, J. P. 1994. The chick-a-dee call system of the Mexican chickadee. *The Condor,* 96, 70–82.

Fitzgibbon, C. D., and Fanshaw, J. H. 1988. Stotting in Thomson's gazelles: an honest signal of condition. *Behavioral Ecology and Sociobiology,* 23, 69–74.

Fouts, R. S., Fouts, D. H., and van Cantfort, T. E. 1989. The infant Loulis learns signs from cross-fostered chimpanzees. In R. A. Gardner, B. T. Gardner, and T. E. van Cantfort (eds.), *Teaching Sign Language to Chimpanzees,* pp. 280–292. State University of New York Press, Albany.

Fox, M. W. 1971. *Behaviour of Wolves, Dogs, and Related Canids.* Jonathan Cape, London.

Fridlund, A. J. 1994. *Human Facial Expression: An Evolutionary View.* Academic Press, San Diego.

Friedmann, H. 1955. The honey guides. *U.S. National Museum Bulletin,* 208, 1–279.

Frings, H., and Frings, M. 1956. Auditory and visual mechanisms in food-finding behaviour of the herring gull. *Wilson Bulletin,* 67, 155–170.

Gardner, R. A., Chiarelli, A. B., Gardner, B. T., and Plooij, F. X. (eds.). 1994. *The Ethological Roots of Culture.* Kluwer Academic Press, Norwell, Mass.

Gardner, R. A., Gardner, B. T., and van Cantfort, T. E. (eds.). 1989. *Teaching Sign Language to Chimpanzees.* State University of New York Press, Albany.

Gautier, J.-P., and Gautier, A. 1977. Communication in Old World monkeys. In T. A. Sebeok (ed.), *How Animals Communicate,* pp. 890–964. Indiana University Press, Bloomington.

Green, S. 1975. Dialects in Japanese monkeys: vocal learning and cultural transmission of locale-specific vocal behaviour. *Zeitschrift für Tierpsychologie,* 38, 304–314.

Greenewalt, C. 1968. *Bird song: Physiology and Acoustics.* Smithonian Institution Press, Washington, D.C.

Gregory, R., and Hopkins, P. 1974. Pupils of a talking parrot. *Nature,* 252, 637–638.

Griffin, D. R. 1958. *Listening in the Dark.* Yale University Press, New Haven.

Grubb, T. C. 1972. Smell and foraging in shearwaters and petrels. *Nature,* 237, 404–405.

Guilford, T., and Dawkins, M. S. 1991. Receiver psychology and the evolution of animal signals. *Animal Behaviour,* 42, 1–14.

Guttmann, G., Predovic, M., and Zemanek, M. 1985. The influence of pet ownership on non-verbal communication and social competence in children. In *Proceedings of the International Symposium on Human-Pet Relationships,* pp. 58–63. IEMT, Vienna.

Gyger, M., and Marler, P. 1988. Food calling in the domestic fowl, *Gallus gallus:* the role of external referents and deception. *Animal Behaviour,* 36, 358–365.

Hanggi, E. B., and Schusterman, R. J. 1994. Underwater acoustic displays and individual variation in male harbour seals, *Phoca vitulina. Animal Behaviour,* 48(6), 1275–1283.

Hare, J. F. 1998. Juvenile Richardson's ground squirrels, *Spermophilus richardsonii,* discriminate among individual alarm callers. *Animal Behaviour,* 55, 451–460.

Harriman, A. E., and Berger, R. H. 1986. Olfactory acuity in the common raven *(Corvus corax). Physiology and Behavior,* 36, 257–262.

Hart, L. A., Hart, B., and Bergin, B. 1987. Socializing effects of service dogs for people with disabilities. *Anthrozoös,* 1, 41.

Hasselquist, D., Bensch, S., and von Schantz, T. 1996. Correlation between male song repertoire, extra-pair paternity, and offspring survival in the great reed warbler. *Nature,* 381, 229–232.

Hausberger, M., Jenkins, P. F., and Keene, J. 1991. Species-specificity and mimicry in bird song: are they paradoxes? A reevaluation of song mimicry in the European starling. *Behaviour,* 117, 53–81.

Hauser, M. D. 1988. How infant vervet monkeys learn to recognise starling alarm calls: the role of experience. *Behaviour,* 105, 187–201.

———. 1989. Ontogenetic changes in the comprehension and production of vervet monkey *(Cercopithecus aethiops)* vocalisations. *Journal of Comparative Psychology,* 103, 149–158.

———. 1993. Right hemisphere dominance for the production of facial expression in monkeys. *Science,* 261(23 July), 475–477.

———. 1996. *The Evolution of Communication.* MIT Press, Cambridge, Mass.

Hauser, M. D., and Andersson, K. 1994. Left hemisphere dominance for processing vocalizations in adult, but not infant, rhesus monkeys: Field experiments. *Proceedings of the National Academy of Sciences,* 91, 3946–3948.

Hauser, M. D., Teixidor, P., Fields, L., and Flaherty, R. 1993. Food-elicited calls in chimpanzees: effects of food quantity and divisibility. *Animal Behaviour,* 45(4), 817–819.

Herman, L. M., Pack, A. A., and Palmer, M-S. 1993. Representational and conceptual skills of dolphins. In H. L. Roitblat, L. M. Herman, and P. E. Nachtigall (eds.), *Language and Communication: Comparative Perspectives,* pp. 403–442. Lawrence Erlbaum, Hillsdale, N.J.

Hess, E. H. 1965. Attitude and pupil size. *Scientific American,* April, 46–54.

Hewes, G. W. 1973. Primate communication and the gestural origin of language. *Current Anthropology,* 14, 5–24.

Hinde, R. A. (ed.). 1969. *Bird Vocalizations: Their Relations to Current Problems in Biology and Psychology.* Cambridge University Press, Cambridge.

Hodun, A., Snowdon, C. T., and Soini, P. 1981. Subspecific variation in the long calls of the tamarin, *Saguinus fuscicollis. Zeitschrift für Tierpsychologie,* 57, 97–110.

Hudson, R., and Vodermayer, T. 1992. Spontaneous and odour-induced chin marking in domestic female rabbits. *Animal Behaviour,* 43(2), 329–336.

Hurst, J. L., Fang, J., and Barnard, C. J. 1993. The role of substrate odours in maintaining social tolerance between male house mice, *Mus musculus domesticus. Animal Behaviour,* 45(5), 997–1006.

Huxley, J. S. 1914. The courtship habits of the great crested grebe *(Podiceps cristatus);* with an addition on the theory of sexual selection. *Proceedings of the Zoological Society of London,* 35, 491–562.

Ingold, T. 1994. From trust to domination: an alternative history of human-animal relations. In A. Manning and J. Serpell (eds.), *Animals and Human Society,* pp. 1–22. Routledge, London.

Jacobs, G. H. 1993. The distribution and nature of colour vision among the mammals. *Biological Reviews,* 68, 413–471.

Janik, V. M., and Slater, P. J. B. 1997. Vocal learning in mammals. *Advances in the Study of Behavior,* 26, 59–99.

Janik, V. M., Dehnhardt, G., and Todt, D. 1994. Signature whistle variations in a bottlenosed dolphin, *Tursiops truncatus. Behavioural Ecology and Sociobiology,* 35, 243–248.

Johnston, R. E., and Jernigan, P. 1994. Golden hamsters recognize individuals, not just individual scents. *Animal Behaviour,* 48(1), 129–136.

Johnston, R. E., Munver, R., and Tung, C. 1995. Scent counter marks: selective memory for the top scent by golden hamsters. *Animal Behaviour,* 49(6), 1435–1442.

Jolly, A. 1966. *Lemur Behavior.* University of Chicago Press, Chicago.

Kaplan, G. 1996. Selective learning and retention: song development and mimicry in the Australian magpie. *International Journal of Psychology. Abstracts of the Twenty-Sixth International Congress of Psychology, Montreal, Canada, 16–21 August 1996,* vol. 31, nos.3 and 4 ("Animal cognition and perception"), p. 233.

———. 1999. Song structure and function of mimicry in the Australian magpie *(Gymnorhina tibicen)* and the lyrebird *(Menura). Advances in Ethology,* vol. 34, ed. Shakunthala Sridhara, supplement b, p. 22.

Kaplan, G., and Rogers, L. J. 1995. Of human fear and indifference: the plight of the orang-utan. In R. D. Nader, B. Galdikas, L. Sheehan, and N. Rosen (eds.), *The Neglected Ape,* pp. 3–12. Plenum Press, New York.

———. 1996. Gaze direction and visual attention in orang-utans. Paper delivered at the Sixteenth Congress of the International Primatological Society, Madison, Wisc.

———. 1999. The *Orang-utans.* Allen and Unwin, Sydney. Also available in 2000 from Perseus Press, Boston.

Karakashian, S. J., Gyger, M., and Marler, P. 1988. Audience effects on alarm calling in chickens *(Gallus gallus). Journal of Comparative Psychology,* 102, 129–135.

Kellert, S. R. 1994. Attitudes, knowledge, and behaviour toward wildlife among the industrial superpowers—the United States, Japan, and Germany. In A. Manning and J. Serpell (eds.), *Animals and Human Society,* pp. 166–187. Routledge, London.

Kellogg, W. N. 1961. *Porpoises and Sonar.* University of Chicago Press, Chicago.

Kilner, R. M., Noble, D. G., and Davies, N. B. 1999. Signals of need in parent-offspring communication and their exploitation by the common cuckoo. *Nature,* 397, 667–672.

Kirn, J. R., Clower, R. P., Kroodsma, D. E., and De Voogd, T. J. 1989. Song-related brain regions in the red-winged blackbird are affected by sex and season but not repertoire size. *Journal of Neurobiology,* 20, 139–163.

Kluender, K. R., Diehl, R. L., and Killeen, P. R. 1987. Japanese quail can learn phonetic categories. *Science,* September, 1195–1197.

Krebs, J. R. 1977. Song and territory in the great tit *Parus major.* In B. Stonehouse and C. Perrins (eds.), *Evolutionary Ecology,* pp. 47–62. Macmillan, London.

Krebs, J. R., and Davies, N. B. 1993. *An Introduction to Behavioural Ecology.* Blackwell Science, Oxford.

Krebs, J. R., Ashcroft, R., and Webber, M. 1978. Song repertoires and territory defence in the great tit. *Nature,* 271, 539–542.

Kroodsma, D. E. 1978. Aspects of learning in the ontogeny of bird song. In G. M. Burghardt and M. Bekoff (eds.), *The Development of Behavior: Comparative and Evolutionary Aspects,* pp. 215–230. Garland, New York.

———. 1990. Using appropriate experimental designs for intended hypotheses in "song" playbacks, with examples of testing effects of song repertoire sizes. *Animal Behaviour,* 40, 1138–1150.

———. 1996. Ecology of passerine song development. In D. E. Kroodsma and E. H. Miller (eds.), *Ecology and Evolution of Acoustic Communication in Birds,* pp. 3–19. Cornell University Press, Ithaca and London,

Kroodsma, D. E., and Parker, L. D. 1977. Vocal virtuosity in the brown thrasher. *Auk,* 94, 783–784.

Kuhl, P. K. 1988. Auditory perception and the evolution of speech. *Human Evolution,* 3, 19–43.

Lambrechts, M. M., Clemmons, J. R., and Hailman, J. P. 1993. Wing quivering of black-capped chickadees with nestlings: invitation or appeasement? *Animal Behaviour,* 46(2), 397–399.

Leal, M., and Rodriguez, J. A. 1997. Signalling displays during predator-prey interactions in a Puerto Rican anole, *Anolis cristatellus. Animal Behaviour,* 54, 1147–1154.

Le Boeuf, B., and Peterson, R. S. 1969. Dialects in elephant seals. *Science,* 166, 1654–1656.

Lehrman, D. S. 1965. Interaction between hormonal and external environments in the regulation of the reproductive cycle of the ring dove. In F. A. Beach (ed.), *Sex and Behaviour.* Wiley and Sons, New York.

Lehtonen, L. 1983. The changing song patterns of the great tit, *Parus major. Ornis Fennica,* 60, 16–21.

Leslie, R. F. 1985. *Lorenzo the Magnificent: The Story of an Orphaned Blue Jay.* John Curley & Associates, South Yarmouth, Mass.

Levinson, B. M. 1969. *Pet-oriented Child Psychotherapy.* Charles C. Thomas, Springfield, Ill.

Lewin, R. 1991, Look who's talking now. Do apes hold the key to the origin of human language? Ape-language studies with one chimpanzee suggest they just might. *New Scientist,* April 27, 39–42.

Lieblich, A. K., Symmes, D., Newman, J. D., and Shapiro, M. 1980. Development of the isolation peep in laboratory-bred squirrel monkeys. *Animal Behaviour,* 28, 1–9.

Lilly, J. C. 1965. Vocal mimicry in *Tursiops:* ability to match numbers and durations of human vocal bursts. *Science,* 147, 300–301.

Lorenz, K. 1941. Vergleichende Bewegungsstudien an Anatiden. *Supplement of the Journal of Ornithology,* 89, 194–294.

———. 1965. *Evolution and Modification of Behavior.* University of Chicago Press, Chicago.

———. 1966. *King Solomon's Ring.* Methuen, London.

Macedonia, J. M., and Evans, C. S. 1993. Variation among mammalian alarm call systems and the problem of meaning in animal signals. *Ethology,* 93, 177–197.

Macedonia, J. M., Evans, C. S., and Losos, J. B. 1994. Male *Anolis* lizards discriminate video-recorded conspecific and heterospecific displays. *Animal Behaviour,* 47, 1220–1223.

MacKinnon, J. 1974. *In Search of the Red Ape.* William Collins, London.

Maclean, C. 1977. *The Wolf Children.* Allen Lane, London.

Malakoff, D. 1999. Following the scent of avian olfaction. *Science,* 286, 704–705.

Mann, N. I., and Slater, P. J. B. 1994. What causes young male zebra finches, *Taeniopygia guttata,* to choose their father as song tutor? *Animal Behaviour,* 47, 671–677.

Manning, A., and Serpell, J. 1994. *Animals and Human Society: Changing Perspectives.* Routledge, London.

Marler, P. 1955, Characteristics of some animal calls. *Nature,* 176, 6–8.

———. 1968. Aggregation and dispersal: two functions in primate communication. In P. C. Jay (ed.), *Primates: Studies in Adaptation and Variability,* pp. 420–438. Holt, Rinehart & Winston, New York.

———. 1970. Bird song and speech development: could there be parallels? *American Science,* 58, 669–673.

———. 1981. Bird song: the acquisition of a learned motor skill. *Trends in Neuroscience,* 4, 88–94.

1991. Differences in behavioural development in closely related species: birdsong. In P. Bateson (ed.), *The Development and Integration of Behaviour,* pp. 41–70. Cambridge University Press, Cambridge.

———. 1997. Three models of song learning: evidence from behaviour. *Journal of Neurobiology,* 33, 501–516.

Marler, P., and Evans, C. 1996. Bird calls: just emotional displays or something more? *Ibis,* 138, 26–33.

Marler, P., and Tamura, M. 1964. Culturally transmitted patterns of vocal behavior in sparrows. *Science,* 146(3650), 1483–1486.

Marler, P., and Tenaza, R. 1977. Signaling behavior of apes with special reference to vocalizations. In T. A. Sebeok (ed.), *How Animals Communicate,* pp. 965–1033. Indiana University Press, Bloomington.

Marshall, A. J., Wrangham, R., Arcadi, A. C. 1999. Does learning affect the structure of vocalizations in chimpanzees? *Animal Behaviour,* 58, 825–830.

May, B., Moody, D. B., and Stebbins, W. C. 1989. Categorical perception of conspecific communication sounds by Japanese macaques, *Macaca fuscata. Journal of the Acoustic Society of America,* 85, 837–847.

Mayford, M., Bach, M. E., Huang, Y-Y, Wang, L., Hawkins, R. D. and Kandel, E. R. 1996. Control of memory formation through regulated expression of a CaMKII transgene. *Science,* 274, 1678–1683.

McComb, K. E. 1991. Female choice for high roaring rates in red deer, *Cervus elaphus. Animal Behaviour,* 41(1), 79–88.

McCracken, G. F. 1993. Locational memory and female-pup reunions in Mexican free-tailed bat maternity colonies. *Animal Behaviour,* 45(4), 811–813.

McFarland, D. 1985. *Animal Behaviour: Psychobiology, Ethology, and Evolution.* Pitman, London.

Mitani, J. C., and Brandt, K. L. 1994. Social factors influence the acoustic variability in the long-distance calls of male chimpanzees. *Ethology,* 96, 233–252.

Mollon, J. D. 1990. Uses and evolutionary origins of primate colour vision. In J. R. Cronly-Dillon and R. L. Gregory (eds.), *Evolution of the Eye and Visual System,* pp. 306–319. Macmillan, New York.

Morrice, M. G., Burton, H. R., and Green, K. 1994. Microgeographic variation and songs in the underwater vocalization repertoire of the Weddell seal *(Leptonychotes weddellii)* from the Vestfold Hills, Antarctica. *Polar Biology,* 14, 441–446.

Morris, D. 1957. Typical intensity and its relation to the problem of ritualisation. *Behaviour,* 11, 1–12.

Narins, P. M. 1990. Seismic communication in anuran amphibians. *BioScience,* 40, 268–274.

Nelsen, A. 1987. *History of German Forestry: Implications for American Wildlife Management.* Yale School of Forestry and Environmental Studies, New Haven.

Neumeyer, C. 1990. The evolution of colour vision. In J. R. Cronly-Dillon and R. L. Gregory (eds.), *Evolution of the Eye and Visual System,* pp. 284–305. Macmillan, New York.

Newby, J. 1997. *The Pact for Survival: Humans and Their Animal Companions.* Australian Broadcasting Corporation, Sydney.

Newman, J. D. 1985. Squirrel monkey communication. In L. A. Rosenblum and C. L. Coe (eds.), *Handbook of Squirrel Monkey Research,* pp. 99–126. Plenum Press, New York.

———. 1995. Vocal ontogeny in macaques and marmosets: convergent and divergent lines of development. In E. Zimmermen, J. D. Newman, and U. Jürgens (eds.), *Current Topics in Primate Vocal Communication,* pp. 73–97. Plenum Press, New York,

Newman, J. D., and Symmes, D. 1982. Inheritance and experience in the acquisition of primate acoustic behavior. In C. T. Snowdon, C. H. Brown, and M. R. Petersen (eds.), *Primate Communication,* pp. 259–278. Cambridge University Press, Cambridge.

Nottebohm, F. 1989. From bird song to neurogenesis. *Scientific American,* February, 55–61.

———. 1992. The origins of vocal learning. *The American Naturalist,* 106, 116–140.

Nottebohm, F., Alvarez-Buylla, A., Cynx, J., Kirn, J., Ling, C.-Y., Nottebohm, M., Suter, R., Tolles, A., and Williams, H. 1990. Song learning in birds: the relation between perception and production *Philosophical Transactions of the Royal Society of London,* 390, 115–124.

O'Connell, S., and Cowlishaw, G. 1994. Infanticide avoidance, sperm competition, and mate choice: the function of copulation calls in female baboons. *Animal Behaviour,* 48(3), 687–694.

Oda, R., and Masataka, N. 1996. Interspecies responses of ringtailed lemurs to playback of antipredator alarm calls given by Verreaux sifakas. *Ethology,* 102, 441–453.

Parr, L. A., and de Waal, F. B. M. 1999. Visual kin recognition in chimpanzees. *Nature,* 399, 647–648.

Partan, S., and Marler, P. 1999. Communication goes multimodal. *Science,* 283, 1272–1273.

Patterson, D. K., and Pepperberg, I. M. 1996. A comparative study of human and parrot phonation: acoustic and articulatory correlates of vowels. *Journal of the Acoustical Society of America,* 96, 634–648.

Payne, R. 1995. *Among Whales.* Scribner, London.

Pepperberg, I. M. 1990a. Cognition in an African Gray parrot *(Psittacus erithacus):* further evidence for comprehension of categories and labels. *Journal of Comparative Psychology,* 104, 41–52.

———. 1990b. Some cognitive capacities of an African Grey parrot *(Psittacus erithacus). Advances in the Study of Behavior,* 19, 357–409.

Pepperberg, I. M., Brese, K. J., and Harris, B. J. 1991. Solitary sound play during acquisition of English vocalisations by an African Grey parrot *(Psittacus erithacus):* possible parallels with children's monologue speech. *Applied Psycholinguistics,* 12, 1151–1178.

Petitto, L. A., and Marentette, P. F. 1991. Babbling in the manual mode: evidence for the ontogeny of language. *Science,* 251, 1493–1496.

Petrie, M., Halliday, T., and Sanders, C. 1991. Peahens prefer peacocks with elaborate trains. *Animal Behaviour,* 41, 323–331.

Pinker, S. 1994. *The Language Instinct.* Penguin Books, London.

Premack, D. 1975. On the origins of language. In M. S. Gazzaniga and C. Blakemore (eds.), *Handbook of Psychobiology,* pp. 591–605. Academic Press, New York.

Preuschoft, S. 1992. Laughter and "smile" in Barbary macaques *(Macaca sylvanus). Ethology,* 91, 220–236.

Provine, R. R. 1996a. Laughter: the study of laughter provides a novel approach to the mechanisms and evolution of vocal production, perception, and social behavior. *American Scientist,* 84 (January/February), 38–45.

———. 1996b. Contagious yawning and laughter: significance for sensory feature detection, motor patterns generation, imitation, and the evolution of social behavior. In C. M. Heyes and B. G. Galef (eds.), *Social Learning in Animals: The Roots of Culture,* pp. 179–208. Academic Press.London.

Putney, R. T. 1985. Do wilful apes know what they are aiming at? *The Psychological Record,* 35, 49–62.

Ralls, K., Fiorelli, P., and Gish, S. 1985. Vocalizations and vocal mimicry in captive harbor seals, *Phoca vitulina. Canadian Journal of Zoology,* 63, 1050–1056.

Randall, J. A. 1994. Discrimination of foot-drumming signatures by kangaroo rats, *Dipodomys spectabilis. Animal Behaviour,* 47(1), 45–54.

———. 1997. Species-specific foot-drumming in kangaroo rats: *Dipodomys ingens, D. desertii, D. spectabilis. Animal Behaviour,* 54, 1167–1175.

Reiss, D., and McCowan, B. M. 1993. Spontaneous vocal mimicry and production by bottlenose dolphins *(Tursiops truncatus):* evidence for vocal learning. *Journal of Comparative Psychology,* 107, 301–312.

Rendall, D., Rodman, P. S., and Edmond, R. E. 1996. Vocal recognition of individuals and kin in free-ranging rhesus monkeys. *Animal Behaviour,* 51(5), 1007–1015.

Robinson, F. N. 1991. Phatic communication in bird song. *Emu,* 91, 61–63.

Robinson, F. N., and Curtis, H. S. 1996. The vocal displays of the lyrebirds *(Menuridae). Emu,* 96, 258–275.

Robinson, F. N., and Robinson, A. 1970. Regional variation in the visual and acoustic signals of the male musk duck *(Biziura lobata). CSIRO Wildlife Research,* 15, 73–78.

Rogers, L. J. 1990. Light input and reversal of functional lateralization in the chicken brain. *Behavioural Brain Research,* 38, 211–221.

———. 1995. *The Development of Brain and Behaviour in the Chicken.* CAB International, Wallingford, U.K.

———. 1996, Behavioral, structural, and neurochemical asymmetries in the avian brain: a model system for studying visual development and processing. *Neuroscience and Biobehavioral Reviews,* 20, 487–503.

———. 1997a. Early experiential effects on laterality: research on chicks has relevance to other species. *Laterality,* 2, 199–219.

———. 1997b. *Minds of Their Own: Thinking and Awareness in Animals.* Allen & Unwin, Sydney, and Westview Press, Boulder.

Roitblat, H. L., Harley, H. E., and Helwig, D. A. 1993. Cognitive processing in artificial language research. In H. L. Roitblat, L. M. Herman, and P. E. Nachtigall (eds.), *Language and Communication: Comparative Perspectives,* pp. 1–23. Lawrence Erlbaum, Hillsdale, N.J.

Roitblat, H. L., Herman, L. M., and Nachtigall, P. E. (eds.). 1993. *Language and Communication: Comparative Perspectives.* Lawrence Erlbaum, Hillsdale, N.J.

Rose, S. 1992. *The Making of Memory.* Bantam, London.

Rumbaugh, D. 1995. Primate language and cognition: common ground. *Social Research,* 62, 711–730.

Ryan, M. J., and Keddy-Hector, A. 1992. Directional patterns of female mate choice and the role of sensory biases. *The American Naturalist,* 139, S4-S35.

Ryan, M. J., and Rand, A. S. 1999. Phylogenetic influence on mating call preferences in female túngara frogs, *Physalaemus pustulosus. Animal Behaviour,* 57, 945–956.

Ryan, M. J., Fox, J. H., Wilczynski, W., and Rand, A. S. 1990. Sexual selection for sensory exploitation in the frog *Physalaemus pustulosus. Nature,* 343, 66–67.

Saetre, G.-P., and Slagsvold, T. 1992. Evidence of sex recognition from plumage colour by the pied flycatcher, *Ficedula hypoleuca. Animal Behaviour,* 44(2), 293–299.

Saito, N., and Maekawa, M. 1993. Birdsong: the interface with human language. *Brain and Development,* 15, 31–40.

Savage-Rumbaugh, S., and Lewin, R. 1994. *Kanzi: The Ape at the Brink of the Human Mind.* Wiley & Sons, New York.

Schaller, G. B. (1963. *The Mountain Gorilla.* University of Chicago Press, Chicago.

Schrader, L., and Todt, D. 1993. Contact call parameters covary with social context in common marmosets, *Callithrix j. jacchus. Animal Behaviour,* 46(5), 1026–1028.

Scott, J. P., and Fuller, J. L. 1965. *Genetics and the Social Behavior of the Dog.* University of Chicago Press, Chicago.

Sebeok, T. A., and Umiker-Sebeok, J. (eds.). 1980. *Speaking of Apes: A Critical Anthology of Two-way Communication with Man.* Plenum Press, New York.

Serpell, J. A. (ed.). 1995. *The Domestic Dog: Its Evolution, Behaviour, and Interactions with People.* Cambridge University Press, Cambridge.

Serpell, J. A. 1996. *In the Company of Animals: A Study of Human-Animal Relationships.* Cambridge University Press, Cambridge.

Settle, R. H., Sommerville, B. A. , McCormick, J., and Broom, D. M. 1994. Human scent matching using specially trained dogs. *Animal Behaviour,* 48(6), 1443–1448.

Seyfarth, R. L., and Cheney, D. M. 1986. Vocal development in vervet monkeys. *Animal Behaviour,* 34, 1640–1658.

Seyfarth, R. M., Cheney, D. L., and Marler, P. 1980. Vervet monkey alarm calls: semantic communication in a free-ranging primate. *Animal Behaviour,* 28, 1070–1094.

Shanas, U., and Terkel, J. 1997. Mole-rat harderian gland secretions inhibit aggression. *Animal Behaviour,* 54, 1255–1263.

Silva, A. J., Paylor, R., Wehner, J. M., and Tonegawa, S. 1992. Impaired spatial learning in alpha-calcium-calmodulin kinase II mutant mice. *Science,* 257, 206–211.

Slater, P. J. B. 1986. The cultural transmission of bird song. *Trends in Ecology and Evolution,* 1, 94–97.

———. 1989. Bird song learning: causes and consequences. *Ethology, Ecology, and Evolution,* 1, 19–46.

Slater, P. J. B., and Ince, S. A. 1979. Cultural evolution in chaffinch song. *Behaviour,* 71, 146–166.

Slater, P. J. B., and Jones, A. E. 1997. Lessons in bird song. *Biologist,* 44, 301–303.

Slater, P. J. B., Ince, S. A., and Colgan, P. W. 1980. Chaffinch song types: their frequencies in the population and distribution between the repertoires of different individuals. *Behaviour,* 75, 207–218.

Slobodchikoff, C. N., Kiriazis, J., Fischer, C., and Creef, E. 1991. Semantic information distinguishing individual predators in the alarm calls of Gunnison's prairie dogs. *Animal Behaviour,* 42, 713–719.

Smith, G. T., Brenowitz, E. A., Beecher, M. D., and Wingfield, J. C. 1997. Seasonal changes in testosterone, neural attributes of song control nuclei, and song structure in wild songbirds. *Journal of Neuroscience,* 17, 6001–6010.

Smith, W. J. 1977. *The Behavior of Communicating: An Ethological Approach.* Harvard University Press, Cambridge, Mass.

Snowdon, C. T. 1993. Linguistic phenomena in the natural communication of animals. In H. L. Roitblat, L. M. Herman, and P. E. Nachtigall (eds.), *Language and Communication: Comparative Perspectives,* pp. 175–194. Lawrence Erlbaum, Hillsdale, N.J.

Stager, K. E. 1967. Avian olfaction. *American Zoologist,* 7, 415–420.

Stebbins, W. C. 1983. *The Acoustic Sense of Animals.* Harvard University Press, Cambridge, Mass.

Stoddart, D. M. 1990. *The Scented Ape.* Cambridge University Press, Cambridge.

Stoddart, D. M., Bradley, A. J., and Hynes, K. L. 1992. Olfactory biology of the marsupial sugar glider—a preliminary study. In R.L. Doty and D. Müller-Schwarze (eds.), *Chemical Signals in Vertebrates,* vol. 6, pp. 523–528. Plenum Press, New York.

Suthers, R. A. 1990. Contributions to birdsong from the left and right sides of the intact syrinx. *Nature,* 347, 473–477.

Thorpe, W. H. 1961. *Bird Song.* Cambridge University Press, Cambridge.

Thorpe, W. H., and Griffin, D. R. 1962, Lack of ultrasonic components in the flight noise of owls. *Nature,* 193, 594–595.

Tinbergen, N. 1953. *Social Behaviour in Animals.* Methuen, London.

———. 1960. *The Herring Gull's World.* Basic Books, New York.

———. 1965. *The Study of Instinct.* Clarendon Press, Oxford.

Todt, D. 1975. Social learning of vocal patterns and modes of their application in grey parrots *(Psittacus erithacus). Zeitschrift für Tierpsychologie,* 39, 178–188.

Todt, D., Goedeking, P., and Symmes, D. (eds.). 1988. *Primate Vocal Communication.* Springer, Berlin.

Trainer, J. M. 1989. Cultural evolution in song dialects of yellow-rumped caciques in Panama. *Ethology,* 80, 190–204.

Tudge, C. 1993. *The Engineer in the Garden: Genes and Genetics from the Idea of Heredity to the Creation of Life.* Jonathan Cape, London.

van Hooff, J. A. 1967. The facial expressions of the catarrhine monkeys and apes. In D. Morris (ed.), *Primate Ethology.* Weidenfeld & Nicolson, London.

Waal, F. de, and Lanting, F. 1997. *Bonobo: The Forgotten Ape.* University of California Press, Berkeley.

Waser, P. M., and Waser, M. S. 1977. Experimental studies of primate vocalization: specializations for long-distance propagation. *Zeitschrift für Tierpsychologie,* 43, 239–263.

Watanabe, S. 1993. Visual and auditory cues in conspecific discrimination learning in Bengalese finches. *Journal of Ethology,* 11, 111–116.

Weary, D., and Krebs, J. 1987. Birds learn song from aggressive tutors. *Nature,* 329, 485.

Wenzel, B. M. 1972. Olfactory sensation in the kiwi and other birds. *Annals of the New York Academy of Sciences,* 188, 183–193.

West, M. J., and King, A. P. 1988. Female visual displays affect the development of male song in the cowbird. *Nature,* 334, 244–246.

West, M. J., King, A. P., and Freeburg, T. M. 1997. Building a social agenda for the study of bird song. In C. T. Snowdon and M. Hausberger (eds.), *Social Influences on Vocal Development,* pp. 41–56. Cambridge University Press, Cambridge.

Wickler, W. 1968. *Mimicry in Plants and Animals.* Weidenfeld & Nicolson, London.

Williams, J. M. and Slater, P. J. B. 1990. Modelling bird song dialects: the influence of repertoire size and numbers of neighbours. *Journal of Theoretical Biology,* 145, 487–496.

Wilson, Edward O. 1975. *Sociobiology: The New Synthesis.* 1975. Harvard University Press, Cambridge, Mass.

———. 1984. *Biophilia.* Harvard University Press, Cambridge, Mass.

Yamagiwa, J. 1992. Functional analysis of social staring behavior in an all-male group of mountain gorillas. *Primates,* 33(4), 523–544.

Zahavi, A. 1975. Mate selection: a selection for a handicap. *Journal of Theoretical Biology,* 53, 205–214.

———. 1979. Ritualisation and the evolution of movement signals. *Behaviour,* 72, 77–81.

Zahavi, A., and Zahavi, A. 1997. *The Handicap Principle.* Oxford University Press, Oxford.

Zann, R. 1990. Song and call learning in wild zebra finches in south-east Australia. *Animal Behaviour,* 40, 811–828.

Zuberbühler, K., Cheney, D. L., and Seyfarth, R. M. 1999. Conceptual semantics in a nonhuman primate. *Journal of Comparative Psychology,* 113(1), 33–42.

INDEX

Andrew J. Niemiec, Ph.D.
Department of Psychology
Kenyon College
Gambier, Ohio 43022